Creative Product Design With Cultural Codes

Ming-Feng Wang

Creative Product Design With Cultural Codes

上海交通大學出版社
SHANGHAI JIAO TONG UNIVERSITY PRESS

Springer

Ming-Feng Wang
General Research Service Center
National Pingtung University of Science
and Technology
Pingtung, Taiwan

ISBN 978-981-19-6766-5 ISBN 978-981-19-6767-2 (eBook)
https://doi.org/10.1007/978-981-19-6767-2

Jointly published with Shanghai Jiao Tong University Press
The print edition is not for sale in the (Mainland) of China. Customers from the (Mainland) of China please order the print book from: Shanghai Jiao Tong University Press.

This Springer imprint is published by the registered company Springer Nature Singapore Pte Ltd.
The registered company address is: 152 Beach Road, #21-01/04 Gateway East, Singapore 189721, Singapore

Preface

Under the circumstances derived by the western cultural policy, Taiwan was impelled to propose the cultural industry in 1995 and continued proposing the cultural intention industry in 2002. It not only explored Taiwan among cultural policies, but economical theory has constant enhancement, and the economic industrial policies have combined closely. According to the civilization and constriction council, in the nearly eleven years from 83 to 93, a run-up in relevant festival celebration of culture activities, it has the tendency to increase gradually in influencing on crowd and consumption capacity. Especially, the degree of consumption of the cultural and creative products is the focus for paying attention to. Because the cultural and creative products are process ways for the local industry and national economy combination, so our country needs to pay attention to this topic even more.

The town Caotun, Nantou County, Taiwan, positively derived local industry to develop the culture industry. Although the people in the town positively tried to develop the local culture industry by holding variety cultural festival celebrations to promote the cultural business opportunity with the accumulated cultures and experience in managing the local industry in decades, it seems unable to reach the goal of culturalization of the products and commercialization of the culture. It was found by investigation that the local cultural and creative products are generally unable to possess the historical, storied, and unique characteristics such that a recognition difference on the cultural and creative products existed between the designer and the consumer and induced problems in getting the cultural merchandise moving. So this book attempts to propose a local cultural and creative products design and development model for raising the distinguishing features of local special products and keeping their economic benefits. Therefore, the aim of this article is to explore how to put the local culture resources in order systematically and collect the local literature and history materials to construct the symbol knowledge base in Caotun with the Semiology theory to offer the designer to the develop product with local culture.

The main method used in the study is Semiology theory method, which can be used to complete the merchandise development model in developing a cultural and creative product, the Scenario-Oriented Design method is also used in this study for designing cultural and creative product exhibitions and strengthening local cultural elements. The results show that the symbol knowledge base is really helpful for developing products having the local culture, history, story, and unique characteristics. It provides the designer a method to develop products based on the obtained local culture, landscape resources, notation, symbolization elements, etc. It not only shortens construction period, but grasps the cultural intention of the products too. This book of study is hoped to be offered to the person who is engaged in the local cultural industry and developed reference on the cultural and creative products.

Kaohsiung, Taiwan Ming-Feng Wang

Acknowledgements

I would like to express my gratitude to Shanghai Summit Discipline in Design, Tongji University Shanghai Institute of Design and Innovation for their financial support to have this book published. I would also like to thank the teachers and students of National Pingtung University of Science and Technology for their encouragement to have this research accomplished.

I owe the completion of this book to my advisor, distinguished professor Ping-Hong Kuo, who has taught me rigorous design, methods, cultural and creative industry, and other research methods, which now serve as the basis for future academic research. In addition to helping guide my writings, he also cultivated my professional abilities in teaching, participating in design competitions, industry-university collaboration projects, and speeches during my studies. I recall that in my spare time, Professor Kuo would take us to engage in a variety of learning and organize initiatives in the community; in life education, he provided students with a constant role model. I am grateful for your endless guidance and support.

I would like to thank Professor Xiao Shiwen of the Department of Industrial Design of National Cheng Kung University, Professor Yang Yufu of the Department of Interior Design of National Yunlin University of Science and Technology, Director Huang Shihui of the Department of Creative Design of National Yunlin University of Science and Technology, and Director Wong Xude of the Department of Industrial Design of Da-Yeh University, for their valuable opinions and contributions to my writing.

The completion of this book would not have been possible without the help of Mr. Xiao Hongcheng for his technical guidance in pottery and celadon. At the same time, I would like to thank Mr. Chu Minsheng, a bamboo art expert, for providing bamboo craft technology and materials. In addition, I would like to thank Huang Shuzhen of the National Taiwan Craft Research and Development Institute for her technical knowledge of straw material and plant dyeing. Thanks to He Yuhui, head of the promotion group of the National Taiwan Craft Research and Development Institute, for her help in the questionnaire and the facilitation of craftsmen introductions. Thanks to Professor Lin Yali from the Department of Statistics of Tunghai University for her guidance on statistical analysis. Thank you, Mama Hongyan, a local senior in Caotun, for an education in the skills of weaving straw sandals; which enabled my successful writing of this book. Thanks to Professor Liu Liwei of National Taiwan University of Arts, for his guidance in curation, which enabled my successful writing of this book. Thanks to the students from the Industrial Design Department at Tunghai University for assisting in arranging the exhibition, so that the design of cultural creative products can be displayed as scheduled. Thanks for the assistance and help of those I was honored to meet throughout the fieldwork, data collection, and in-depth interviews during the writing of the book, enthusiastic cultural figures and groups including: Chen Guoda, former chairman of the Caotun Township Local Culture and Education Association, chairman Chen Zhizhi, manager Xiao Jingfen, in association with many other association members and volunteers.

Thanks to upperclassmen Lin Genghui, Guo Zhenwei, Lai Kuanxian, and Yang Jiafeng, of the Institute of Industrial Design, Tunghai University, for their training in writing, my classmates Wang Zhuanbo and Wu Jiannan for their guidance in writing and statistical analysis, my friends Xu Zhenyao, Zeng Xinyi, Weng Zhiliang, Du Zhehao, Liu Rongxiang, Chen Yuezhong for grammatical revisions, and fellow graduate underclassmen Wu Junxiang, Wu Ruoping, Shi Guangsheng, Liu Jiawei, Zhuang Yusheng, Liu Pinxuan, Tu Zhengqin, Chen Youru, Li Yanliang, college senior Su Zhaowei and Xie Peiling for the arrangement of the venue during the cultural and creative product exhibition, and assistance putting on my exhibition. Thank you.

July 2022 Ming-Feng Wang

Contents

1 Introduction 1
1.1 General Background Information of the Book 1
1.2 Motive of the Book 3
1.3 Objective of the Book 4
1.4 Terminology of the Book 4
1.5 Object and Scope of the Book 7
1.6 Subjects of the Book 9
1.7 Main Research Methods 10
1.8 Framework and Description of the Book 10
References 12

2 Research Methods 13
2.1 Research Methods and Process 13
2.2 Qualitative Research 13
2.3 Quantitative Analysis 19
References 22

3 Literature Review 25
3.1 Field Survey 25
3.2 Domestic and International Development Trends of Cultural and Creative Industries 30
3.3 Definition and Categorization of Cultural and Creative Products 39
3.3.1 Categories of Cultural and Creative Products 41
3.3.2 Analysis of Cultural and Creative Commodities with Monopolistic Consumption Characteristics in Taiwanese Museums, Communities, and Local Areas 45
3.4 Consumers' Perceptions of Cultural and Creative Products 74
3.5 Relevant Domestic and International Research and Positioning 76
References 85

4 Construction of the Knowledge Base Concept Form and Process 89
4.1 Literature on Semiotics 89
4.1.1 Relevant Theories of Various Semiotic Scholars 90
4.1.2 Symbol Conversion and Interpretation 96
4.2 Design Models and Procedures of Cultural and Creative Products 98
4.3 The Contents and Steps for the Construction of Cultural and Creative Products Design and Development Model 99
4.3.1 Step 1: The Original Concept of the Knowledge Base of Cultural and Historical Design Symbols 100
4.3.2 Step II: Build Knowledge Base Horizontal and Vertical Axis Design and Format Content 101
4.3.3 Step III: Select Local Resources 106
4.3.4 Step IV: Complete the Construction of the Knowledge Base Form 114
4.3.5 Step V: Application and Procedure of Knowledge Base 116
4.4 Results of Cultural and Creative Product Design as Viewed in Knowledge Base 118
4.5 Results of Knowledge Base Feasibility Research and Analysis 120
4.5.1 Definition of Terms Related to Statistics 122
4.5.2 Questionnaire 125
4.5.3 Analysis of the Basic Data of the Subjects 126
4.5.4 Knowledge Base Training Appropriate Time Analysis 127
4.5.5 Feasibility Analysis of the Conceptual Form of the Knowledge Base of Cultural and Historical Design Symbols to Assist Designers in Developing Products 128
4.5.6 Analysis of Differences Before and After Using the Knowledge Base 131
4.5.7 Analysis and Comparison of the Ease or Difficulty of the Development Mode of Cultural Creative Products 134
4.5.8 Analysis of Satisfaction of Cultural Creative Product Achievements After Design 136
References 139

5 Cultural and Creative Products Exhibition and Its Analysis Based on Scenario-Oriented Design 141
5.1 Literature on Scenario-Oriented Design 141
5.2 Cultural and Creative Products Exhibition and Its Procedure Based on Scenario-Oriented Design 145
5.3 Scenario Composition and Setting 147
5.3.1 "Initiate" Planning Steps 148
5.3.2 "Inherit" Planning Steps 149
5.3.3 "Transfer" Planning Steps 150

5.3.4 "Integration" Planning Steps ... 150
5.3.5 Scenario Planning of the Culture Creative Commodities Exhibition ... 152
5.4 Space Planning of Cultural and Creative Product Exhibitions ... 153
5.5 Cognitive Appraisal and Analysis of Cultural and Creative Product Exhibitions ... 165
5.5.1 Questionnaire ... 166
5.5.2 Basic Data Analysis of Subjects ... 171
5.5.3 Analysis of Cognition of Caotun Cultural Resource Impression ... 174
5.5.4 Consumption and Analysis of the Cultural Creative Products on Exhibit ... 182
5.5.5 Analysis of the Cognition Towards the Cultural Creative Products Exhibition ... 185
5.5.6 Effect Analysis of Scenario-Based Method ... 190
5.5.7 Definition and Status Snalysis of Local Cultural Creative Products ... 195
References ... 201

6 Comprehensive Discussion ... 203
6.1 Analysis Results and Discussion of Knowledge Base Conceptual Form Establishment and Process ... 203
6.2 Analysis Results and Discussion of the Scenario-Based Cultural Creative Product Exhibition ... 204

7 Conclusions and Recommendations ... 207
7.1 Summary of the Book ... 207
7.2 Recommendations of the Book ... 210

Index ... 213

About the Author

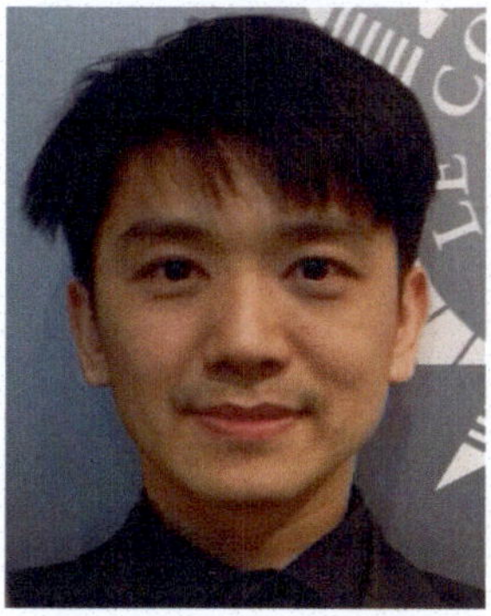

Ming-Feng Wang was selected as one of the Outstanding Young Persons in Taiwan in 2009. He received his Ph.D. degree from the Department of Industrial Design, National Cheng Kung University in 2016, and is currently a Research Fellow and Adjunct Assistant Professor at the National Pingtung University of Science Technology, Taiwan. His research interests include the Cultural and Creative Industry, Product Design, Industrial Design, History, Fine Arts and Crafts. He has published more than 25 SCI, EI and Conference papers, along with three books.

List of Figures

Fig. 1.1	Administrative map of Caotun	8
Fig. 1.2	Research process and core	11
Fig. 2.1	Research methods and steps	14
Fig. 3.1	Correlation map of natural, production, human, cultural and landscape in Caotun	26
Fig. 3.2	The evolution of Taiwan's cultural policy	35
Fig. 3.3	The process from culturalization of straw sandal craft industry to cultural industrialization	36
Fig. 3.4	The difference between cultural and creative products and general products (Ho et al., 1996; Wang, 2020)	40
Fig. 3.5	Starbucks coffee	42
Fig. 3.6	IKEA	42
Fig. 3.7	Calligraphy by Chang, Dai-Chien (Chung, Li-Her Museum 2022)	43
Fig. 3.8	Blue-and-white porcelain (The Shanghai Museum, 2017)	44
Fig. 3.9	Cultural and creative products of the National Palace Museum in Taiwan (The National Palace Museum, 2022)	44
Fig. 3.10	Cultural and creative products of the Yilan BaiMi clogs community (Wooden Clog Museum, 2022)	45
Fig. 3.11	Cultural and creative products of various ethnic groups in Taiwan (Wang, 2018)	45
Fig. 3.12	Cultural and creative products of local health slippers in Caotun, Nantou (Caotun Straw Crafts Museum, 2020)	46
Fig. 3.13	Exhibition hall—main building (The National Palace Museum, 2022)	47
Fig. 3.14	The National Palace Museum (2022)	48
Fig. 3.15	Xi Bodkin with dragon pattern, late warring states period to early Western Han dynasty, circa 275–141 B.C.E. (The National Palace Museum, 2022)	48
Fig. 3.16	Dragon opened the letter opener (The National Palace Museum Shop, 2022)	49

Fig. 3.17 Meat-shaped stone, Qing dynasty, AD1644-1911 (The National Palace Museum, 2022) 49
Fig. 3.18 National treasure book clip: meat-shaped stone (The National Palace Museum Shop, 2022) 49
Fig. 3.19 Pillow in the shape of a recumbent child with white glaze, Ding ware, Northern Song dynasty AD960-1127 (The National Palace Museum, 2022) 50
Fig. 3.20 Paperweight-pillow in the shape of a recumbent child (The National Palace Museum Shop, 2022) 51
Fig. 3.21 Jadeite Cabbage, Qing dynasty, AD1644-1911 (The National Palace Museum, 2022) 52
Fig. 3.22 Jadeite Cabbage digital camera (The National Palace Museum Shop, 2022) 53
Fig. 3.23 Jade Pi-Hsieh, auspicious beast, Eastern Han dynasty AD25-2201 (The National Palace Museum 2022) 53
Fig. 3.24 Paperclip holder paper weight-Pi-hsieh (The National Palace Museum Shop, 2022) 54
Fig. 3.25 White rice village wall (Wooden Clog Museum, 2022) 54
Fig. 3.26 Wooden clog box (Wooden Clog Museum, 2022) 55
Fig. 3.27 Schefflera tree (Wooden Clog Museum, 2022) 55
Fig. 3.28 Wooden clog museum (Wooden Clog Museum, 2022) 56
Fig. 3.29 The 3rd Floor briefing room (Wooden Clog Museum, 2022) 56
Fig. 3.30 The 3rd Floor balcony (Wooden Clog Museum, 2022) 57
Fig. 3.31 The 2nd Floor clog making process exhibit (Wooden Clog Museum, 2022) 58
Fig. 3.32 The 2nd Floor large-scale clog exhibit (Wooden Clog Museum, 2022) 58
Fig. 3.33 The 2nd Floor tandem clog exhibit (Wooden Clog Museum, 2022) 59
Fig. 3.34 The 2nd Floor clog wall display (Wooden Clog Museum, 2022) 60
Fig. 3.35 Wooden clog (Wooden Clog Museum, 2022) 60
Fig. 3.36 The 1st Floor decorative clog exhibit (Wooden Clog Museum, 2022) 61
Fig. 3.37 The 1st Floor Clog making tools exhibit (Wooden Clog Museum, 2022) 62
Fig. 3.38 The 1st Floor DIY tool cabinet/table (Wooden Clog Museum, 2022) 63
Fig. 3.39 The 1st Floor clog display case (Wooden Clog Museum, 2022) 64
Fig. 3.40 The 1st Floor large scale painted clog exhibit (Wooden Clog Museum, 2022) 65
Fig. 3.41 The box can hold candy and nuts (Wang, 2007) 65
Fig. 3.42 The pens box (Wang, 2007) 66
Fig. 3.43 The box can hold other items (Wang, 2007) 66

Fig. 3.44 The tea box (Wang, 2007) 67
Fig. 3.45 Lucky Box for Shansan Kings (Wang, 2007) 67
Fig. 3.46 Hakka wooden cutting board concept sketch (Wang, 2007) 68
Fig. 3.47 Hakka Magnolia Coco wood cutting board (Wang, 2007) 68
Fig. 3.48 The main tools of straw sandal weaving 72
Fig. 3.49 There is a understanding gap between mass consumers and local cultural and creative products 76
Fig. 3.50 The comprehension process of cultural and creative products 77
Fig. 3.51 Solving the comprehension issue among designers and consumers 84
Fig. 4.1 Caotun signifier and Signified 92
Fig. 4.2 Charles Morris' three properties of symbols 93
Fig. 4.3 Roland Barthes' definitions of denotation and connotation 94
Fig. 4.4 Commodity value (Hoshino, 1991) 96
Fig. 4.5 The contents and steps for the construction of cultural and creative products design and development model 99
Fig. 4.6 Transformation of local cultural and historical data into local intellectual assets 101
Fig. 4.7 The three-leveled culture (Yang, 1998) 104
Fig. 4.8 Integrate the three groups of concepts to build a knowledge base in the form of lateral and vertical thinking 105
Fig. 4.9 Divide the content of the knowledge base into sections focused on either vertical or horizontal thinking 106
Fig. 4.10 Input and application of local knowledge base resources 114
Fig. 4.11 Cultural and historical design symbols knowledge base about natural, production, human, cultural and landscape resources 117
Fig. 4.12 The application and process of the knowledge base 118
Fig. 4.13 Knowledge base training time percentage analysis 128
Fig. 4.14 Percentage analysis of whether cultural creative products can be effectively designed and developed after using the knowledge base 130
Fig. 4.15 Feasibility percentage analysis of applying the knowledge base structure in different places 131
Fig. 4.16 Before using the knowledge base 132
Fig. 4.17 After using the knowledge base 132
Fig. 4.18 Bar chart of the most difficult steps in the design process of cultural creative products 134
Fig. 4.19 Easiest step in the cultural creative product design process bar chart 136
Fig. 5.1 The relationship between characters, objects, and scenario 143
Fig. 5.2 Journey map (Yu et al., 2000) 145
Fig. 5.3 Steps for the scenario-based design and planning process for the cultural creative product exhibition 147

Fig. 5.4 The purpose of the Lugang Danpushe 149
Fig. 5.5 Poster for the cultural creative product exhibition 150
Fig. 5.6 Introductive extension poster for each stop 151
Fig. 5.7 Cultural creative product exhibition invitation card front 151
Fig. 5.8 Cultural creative product exhibition invitation card back 152
Fig. 5.9 Cultural creative product exhibition CD cover 153
Fig. 5.10 Life-size Caotun Monopoly game display concept 154
Fig. 5.11 Site layout of the Caotun Straw Craft Culture Center 156
Fig. 5.12 The layout of the exhibition center on the 2F of Caotun Town Office 156
Fig. 5.13 Taichung City Dadun Cultural Center, Wenying building configuration map and site configuration map 157
Fig. 5.14 **a–c** Exhibit 1:20 scale model 157
Fig. 5.15 **a**, **b** The 1st stop: Lugang starting point 159
Fig. 5.16 Tang Thau Khin 160
Fig. 5.17 The 2nd stop: Figurines of ancestral characters 162
Fig. 5.18 Doll of Caotun ancestors 163
Fig. 5.19 The 3rd stop: Dengying academy 165
Fig. 5.20 Dengying book case 166
Fig. 5.21 Pious heart 167
Fig. 5.22 Valedictorian candle tray 167
Fig. 5.23 The 4th stop: Furong hall 169
Fig. 5.24 Xiu Ju Dongnan 170
Fig. 5.25 **a**, **b** 5th stop: Hoards of shoes 171
Fig. 5.26 A mound of shoes 172
Fig. 5.27 Innovative straw sandal weaving and design 173
Fig. 5.28 The 6th stop: Eight-diagrams Kiln 175
Fig. 5.29 Bagua Peak 175
Fig. 5.30 Tea fragrance Tai Chi 176
Fig. 5.31 The 7th stop: Nanpu tobacco barn 178
Fig. 5.32 Yanhai Yunlou 179
Fig. 5.33 The 8th stop: Jiujiu peak 180
Fig. 5.34 Lei Gong strikes the emperor 181
Fig. 5.35 Cao Xie Dun tea house 182
Fig. 5.36 Jiujiu peak 183
Fig. 5.37 The 9th stop: Tombstone Mountain 186
Fig. 5.38 Tombstone phoenix crown 187
Fig. 5.39 **a**, **b** The 10th stop: Puli terminal station 188
Fig. 5.40 Responsibility 189
Fig. 5.41 Analysis chart of the opinions of foreign tourists and local residents towards Caotun's spiritual symbols 192
Fig. 5.42 A pie chart analysis of the origin of Caotun's place names as believed foreign tourists and local residents 193
Fig. 5.43 A pie chart analysis of Caotun resources by foreign tourists and local residents 194

Fig. 5.44 Bar chart analysis of respondents' impression of Caotun's landscape resources 195
Fig. 5.45 A pie chart analysis of the evaluation of the awareness of Caotun local culture by cultural and creative commodity exhibitions 198
Fig. 5.46 Pie analysis chart of respondents' definition and status of cultural and creative products 200
Fig. 6.1 Knowledge base is the bridge between foreign and local designers 204
Fig. 6.2 The respondents' definition of local cultural creative products 206
Fig. 6.3 The respondents' determination of the status of local cultural creative products 206

List of Tables

Table 2.1 Five major local resources 16
Table 2.2 Subjects 17
Table 2.3 Respondents to questionnaires with Close-ended questions 21
Table 2.4 Respondents to questionnaires with open-ended questions 21
Table 3.1 Field survey on natural, production, human, cultural and landscape in Caotun 28
Table 3.2 SWOT analysis on future development of the straw art industry in Caotun (Chen, 2005; Wang, 2007) 29
Table 3.3 Policies on cultural industries in Taiwan (Wang, 2003, 2005) 33
Table 3.4 Policies on creative industries in Taiwan (Wang, 2003, 2005) 34
Table 3.5 Culture and industry: definition and merits/demerits (Lin, 1998) 35
Table 3.6 Statistics on the attendance of national art and cultural events (National Arts & Cultural Activities Information System of the Ministry of Culture, 2022) 39
Table 3.7 Evolution of straw sandals in Caotun 70
Table 3.8 Process of making straw sandals 73
Table 3.9 Creative straw-based merchandise in Caotun (Caotun Straw Crafts Museum, 2005) 75
Table 3.10 Domestic and foreign research on creative product design 79
Table 4.1 Interpreting Caotun's three types of cultural codes 91
Table 4.2 Syntagmatic and paradigmatic analyses of Caotun's culture 93
Table 4.3 Meaning and denotation of Caotun's Dengying academy 95
Table 4.4 Accommodation of assimilation and dissimilation among creative straw-based products 97
Table 4.5 Code conversion of saccharin and straw sandals (Birkhoff, 1973; Hu, 2010; Wang, 2007; Yang, 1998) 98

Table 4.6 Theories proposed by semiotic researchers (Wang, 2020) 103
Table 4.7 Summary table of natural, production, human, cultural and landscape in Caotun 108
Table 4.8 Selection of eight representative resources in Caotun 113
Table 4.9 Jiujiu Peak's cultural and historical design symbols knowledge base 115
Table 4.10 Design and manufacturing processes of creative products 119
Table 4.11 Review on design results of creative products 121
Table 4.12 Specialties of respondents 125
Table 4.13 Training duration and locations of respondents 126
Table 4.14 Cross tabulation analysis: respondents' occupations and basic information 127
Table 4.15 Cross tabulation analysis: duration of training on design thinking process 129
Table 4.16 Feasibility analysis on design model presented in this study 130
Table 4.17 Feasibility analysis of applying design model used in this study in other cases 131
Table 4.18 Statistical analysis on effectiveness of design model: before and after 133
Table 4.19 Most difficult stage of design thinking process 135
Table 4.20 Easiest stage of design thinking process 137
Table 4.21 Satisfactory levels and factor analysis of design results 138
Table 5.1 List of ancestral characters 144
Table 5.2 Map journal (Yu et al., 2000) 144
Table 5.3 Routes taken by ancestors 146
Table 5.4 Design and planning of creative products 155
Table 5.5 Platforms, decorations, and props used in creative products expos 158
Table 5.6 The 1st stop: Lugang starting point 161
Table 5.7 The 2nd stop: Figurines of ancestral characters 164
Table 5.8 The 3rd stop: Dengying academy 168
Table 5.9 The 4th stop: Furong hall 171
Table 5.10 The 5th stop: Hoards of shoes 174
Table 5.11 The 6th stop: Eight-diagrams Kiln 177
Table 5.12 The 7th stop: Nanpu tobacco barn 180
Table 5.13 The 8th stop: Jiujiu peak 184
Table 5.14 The 9th stop: Tombstone Mountain 188
Table 5.15 The 10th stop: Puli terminal station 190
Table 5.16 Locations of creative products expos 190
Table 5.17 Cross tabulation analysis: respondents' basic information 191
Table 5.18 Cross tabulation analysis: impression of Caotun's symbols (locals residents vs. foreign tourists) 192
Table 5.19 Cross tabulation analysis: Caotun's name origin (locals residents vs. foreign tourists) 193

Table 5.20 Cross tabulation analysis: assessment of Caotun's five major resources by locals residents and foreign tourists 195
Table 5.21 Assessment of Caotun's landscape resources by locals residents and foreign tourists 196
Table 5.22 Analysis on purchases of creative merchandise in expos (locals residents vs. foreign tourists) 197
Table 5.23 Feedback on display and planning of creative products from locals residents and foreign tourists 197
Table 5.24 Overall evaluation and analysis of creative products expos 198
Table 5.25 Survey on creative products expos' representation of local cultures 199
Table 5.26 Validity of questionnaires 199
Table 5.27 Cross tabulation analysis: positioning of local creative products by designers and consumers 200

Chapter 1
Introduction

1.1 General Background Information of the Book

First proposed by Australia and the United Kingdom, followed by Canada, New Zealand, the United States, Finland, along with different countries in Europe and Asia, all of which have proposed similar cultural policies, economic discourse has become the most important part of cultural policy (Wang, 2003, 2005). Related policies and industry evaluations are constantly being announced, and each country has unique news and developments regarding the output value brought by the creative industry, the contribution to gross national product, the increase in employment opportunities, and the expansion rate of the market. Cultural policy is economic policy, culture creates wealth, culture adds value and makes an indispensable contribution to innovation, marketing and design. In addition to culture itself being a valuable output, it also has an indispensable added value to the output of other commodities. It can be said that culture plays a pivotal role in economic success (Department of Communication and the Arts, 1994).

In Taiwan, the core concept and policy of "community development" was promoted in the 1960s. The term "community development" was gradually adopted after Chen Qinan advocated a new concept of "community building" at the Council for Cultural Affairs in 1994. The term takes "establishing community culture, building community consensus, and constructing the concept of community life community as a new thinking and policy of cultural administration" as its main goal. Generally speaking, community building is to act on different types of community issues. Japanese Professor Kiyoshi Miyazaki advocates that these issues be divided into five categories: "people", "text", "land", "production" and "Landscape". The main purpose is to integrate the five major community development aspects of "people, culture, land, Landscape, and production", and the policy term is generated (Ministry of Culture, 2008; Su, 2009).

M.-F. Wang, *Creative Product Design With Cultural Codes*,
https://doi.org/10.1007/978-981-19-6767-2_1

From the perspective of the development of Taiwan's cultural policy, the concept of economy and industry was mentioned in the "Culture and Industry", a seminar of the Council of Cultural Affairs in 1995. In the name of "culture industry", it includes local crafts, tourism, historical site preservation, media, art of living, agriculture and fishery, cultural sponsorship of enterprises, etc. (National Culture and Arts Foundation, 2004). Taiwan's cultural policy attempts to "culturalize industry, industrialize culture", initiated by local governments, from small areas to large areas, from point to surface, to create the cultural characteristics of national industries (Weng & Miyazaki, 1996). After 2002, Taiwan implemented the "Challenge 2008 National Development Plan", and "cultural and creative industries" is one of its key points, covering a wide range of traditional culture, culture industry, knowledge economy and surrounding industries.

In recent years, with the development of industry and commerce, the trend of "urbanization" in Taiwan has become more and more obvious. A large number of employment opportunities and youth population are concentrated in urban areas. According to statistics collected by the Ministry of the Interior, migrant populations in the six major Taiwanese have held a steady majority over the years. In 2016, the migrant population in the six major (Taipei City, New Taipei City, Taoyuan City, Taichung City, Tainan City, Kaohsiung City) cities accounted for 72% of the national population. In the past 7 years, sorted by region, the northern region of the country has the largest net immigrant population, indicating that the population continues to concentrate in the north, and the issue of regional balance deserves attention (Executive Yuan, 2022).

As a result of over-urbanization, rural areas suffer from population reduction and aging, along with the issue of unbalanced urban–rural development. In order to solve the aforementioned problems, the government refers to Japan's experience in promoting "local creation", and has promoted the "design flip, local creation" plan since 2016. Through taking advantage of local existing "land, production, and people" resources, the plan aims to reinforce the uniqueness and core values of each locality, and the introduction of the guidance mechanism of "creativity (design power), innovation (productivity), and entrepreneurship (marketing power)", inject industrial development momentum into the locality. As a pilot demonstration project, after being promoted from the perspective of "design power", not only has the initiative successfully established demonstration cases, but also has attracted attention and responses from all sectors of society and the media on local creation issues through the holding of achievement exhibitions. The next stage will be to fully integrate the resources of various government ministries, encourage enterprises to invest in their hometowns, and introduce technology and knowledge into local areas to improve productivity and marketing power, create opportunities for local industrial innovation, improve local living environments, and attract local youths to return with the ultimate goal of balancing rural and urban populations in Taiwan (Executive Yuan, 2022).

All local communities and rural areas in Taiwan have unique cultural connotations, landscapes, industrial history, and handicraft inheritance. In order to assist local governments in rediscovering the cultural heritage of their communities, the

National Development Council began promoting the "design flip, local creation" plan in 2016. By taking stock of the characteristic resources of "land, production, and people" in various places, with the strategic planning of "creativity, innovation, entrepreneurship, and creation", the plan aims to develop valuable local industrial resources while guiding talent to give back to their hometowns through providing professional services. Through the diversified combination of regions, industries and local talent, and the value-added application of design methods, the plan will promote industrial development and local cultural improvement, enabling local communities, settlements, and villages to usher in a modern golden age, one which exhibits the beauty of the landscape and redefines the place Identity (National Development Council, 2022).

The promotion of Taiwan's "culture industries" and "cultural and creative industries" has enriched local culture industries with creativity. At the same time, the policies of "community development" and "local creation" are based on building community culture, creating community consensus, and constructing a community of life to increase productivity and marketing power. These policies achieve the goal of creating opportunities for local industry innovation, improving the local living environment, attracting local youth to return to their hometowns, and balancing the population distribution in Taiwan. The purpose of this book is to organize and summarize the transformation of local traditional industries, so that "industrial culturalization and cultural industrialization" can be better implemented in research, and to provide opinions and ideas on the areas that need to be given attention in Taiwan's national plan. At the same time, the book uses traditional industries in the Caotun township as a research basis to identify challenges faced in these industries and propose solutions, with the aim of being helpful to future cultural and creative industries in Taiwan.

1.2 Motive of the Book

Taiwan's cultural industry policy have begun to take shape and developed well since it was promoted, but in the process of carrying out the policy of "culture industrialization and industry culturalization", the complete transformation hasn't been accomplished. Take Caotun Township of Nantou County as an example, it holds "to build the county by craftsmanship" as its long-term goal, and drives the economic benefit of local cultural industry. "The international cultural festival of straw craftsmanship" has been held for many years, and with the promotion of the festival, the local cultural and creative products have undergone innovation and transformation, which strengthened the consumers' deep understanding of the unique local culture, helped the local industry, and broadened the consumption vision. However, the local craftsmen know well of the cultural history background, and have much experience and advanced craft, and they are engaged in developing the local cultural and creative products, but they do not master the symbolic meaning and cultural spirit of the goods, which become the bottleneck of the breakthrough in the process of "cultural commercialization".

In this study, local resources, including land, people, culture and the Landscape, were collected and developed into database, and a development model for local goods was proposed in the hope that the cultural cognitive difference would be reduced in terms of the design resources, and the value and implication of the cultural and creative products would be promoted. The study adopted "the cultural and creative products of Caotun Township, Nantou County" as the research objective, and carried out analyses and discussions, pointing out how the local cultural and creative products of Caotun Township, Nantou County can find its footing in the market. The research purpose is to construct a knowledge base usage pattern from the theory of semiotics, provide a development foundation for related goods designers, and reduce the cognitive difference of cultural and creative products between the designers in order to help the design and promotion of local cultural and creative product.

1.3 Objective of the Book

The result of this research is the analysis and discussion of "The Cultural and Creative Products of Caotun Township, Nantou County", as well as the identification of the foothold of local cultural creative commodities, using "Cultural Creative Products of Caotun Township, Nantou County" as the research subject. The project has the following two goals:

1. This research uses semiotics theory to construct a knowledge base usage model, providing local and foreign designers as the basis for product development, shortening the cognitive gap between local and foreign designers in cultural and creative commodities, and facilitates the design and promotion of local cultural and creative products.
2. This research integrates the scenario-oriented design method along with cultural and creative products to exhibit evidence, create a roadmap, and tell stories, so that the unique stories of each locality are sequentially presented, thus strengthening the cultural understanding of tourists and local residents.

1.4 Terminology of the Book

The definition of culture is broad. In order to avoid the dilemma of ambiguity and confusion in presentation of research, the following section will define and explain all relevant terms, including Culture, Culture Industries, Culture Creative Industries, Qualitative Research, Quantitative Analysis, Local Cultural and Creative Products, Cultural Concreteness, Cultural Connotation, Cultural and Historical Design Symbols Knowledge Base, Caotun (Cao Xie Dun) and Lugang Danpushe, etc.

1. Culture

British anthropologist Edward Burnett Tylor (1832–1917) once defined "culture" in a broad sense, stating: "Culture or civilization, in its broad national sense, is the complex whole which includes knowledge, belief, art, law, morals, custom, and any other capabilities and habits acquired by man as a member of society". Over time, this cultural definition has become the classic definition of "culture". In other words; culture is everything that is closely related to people's social collective life (Gao, 2002).

2. Culture industries

According to the United Nations Educational, Scientific and Cultural Organization (UNESCO), the definition of *culture industries* is the industries, which combine the creation, production and commercialization of intangible and culturally-essential content. Such content is typically protected by copyright and may be represented in the form of a product or service. The concept of cultural industries can include printing, publishing, multimedia, audio and visual, photography and film production, and craft and design (Chen, 2005; Lin, 2011; Council for Cultural Affairs, Executive Yuan, 2022). Taiwan subsequently proposed the term "culture industry" in 1995, which belongs to the traditional economy and is what we call community development.

3. Culture creative industries

The term Cultural Creative Industries is what Taiwan uses to describe the international culture industry or creative industry. In 2002, the "Challenge 2008—National Development Plan" first proposed the term "cultural and creative Industries", and actively used the term in reference to the conceptual form of the industrial chain, redefining the value of the culture industry, in order to open up the creative field, and combine the humanities and the economy to develop an industry that takes into account cultural accumulation and economic benefits. Generally speaking, there is no substantial difference in international studies between "cultural and creative industries" and the "cultural industry" or "creative industry (Chen, 2005)."

4. Qualitative research

Qualitative Research refers to the use of non-mathematical analysis procedures, through various methods, including observation, interviews, etc. to collect data and produce research results. Mainly used in this study to construct the cultural meaning of social reality, focusing on the process of interactions. In Qualitative Research, authenticity is key, and its value is ubiquitous and obvious. Studies are limited by context, and contain a few test subjects, results are analyzed using thematic analysis, and requires the researcher to be heavily involved (Neuman, 2005).

5. Quantitative analysis

Quantitative research is defined by its focus on measuring objective facts, its focus on variables, the importance of reliability and repeatability, must be unaffected by context, studies contain multiple case subjects, results are analyzed with statistical analysis, and the researcher maintains a neutral stance (Neuman, 2005). This study uses the questionnaire format, and uses the SPSS statistics software to aggregate the related data.

6. Local culture industries

The word "local" emphasizes the locality. No matter domestic or foreign, the natural environment, cultural background, and cultural customs of each location are different, and the products formed have their own uniqueness. Because of these distinctive natural conditions, distinctive historical background, distinctive people, and distinctive knowledge, these products with distinctive significance can be presented (Ming, 2000). Thus these products are cultural, historical, indicative of place, representative, symbolic, and irreplaceable. Intrinsic development is to take the location itself as the starting point of analytical thought, and develop industries based on local characteristics, local conditions, local talents, and even the local well-being (Chen, 1995a, b).

7. Cultural and creative products design and development model

Based on semiotic theory, this book constructs the "Form of Cultural and Historical Design Symbol Knowledge Base", which is the so-called cultural and creative product development model, mainly for use by designers.

8. Local cultural and creative products

Commodities derived from local cultural characteristics have local significance and economic value. They mainly exist to bring economic benefits to the locality. Consumers need to obtain commodities in the way of pilgrimage, forming a local monopoly-style marketing category of cultural and creative commodities. Due to this, these products are known as "local cultural and creative products".

9. Cultural concreteness

In the theory of semiotics, it is believed that symbols have clear surface meanings, similar to the "extended meaning" of cultural and creative commodities, which refers to the tangible substances of local cultural and creative commodities, such as shape, color, texture, function, etc., hereinafter referred to as "cultural concreteness".

10. Cultural connotation

Symbols have hidden deep meanings, just like the connotation of cultural and creative commodities, and their meanings are expressed in the invisible spirit behind the locality, such as local stories, symbols and history, hereinafter referred to as "cultural connotation".

11. Cultural and historical design symbols knowledge base

According to the original concept of knowledge base in enterprise management, this paper innovatively introduces the local cultural and creative commodity development mode to solve the needs of related commodity design and developers (hereinafter referred to as the knowledge base).

12. Caotun Township (Cao Xie Dun)

Caotun Town, the largest township in Taiwan, formerly known as "Cao Xie Dun" (Straw Sandal Village), is storied to have gotten its name during the Qianlong and Jiaqing periods. The settlers, porters, and business travelers who traveled from Lukang to Puli arrived there and discarded their old, worn-out straw sandals, replacing them with new ones. Over time, the village accumulated a large number of old straw sandals ("Cao Xie Cheng Dun"), eventually the name became simply "Caotun", the timing and historical records give this accounting the most credibility.

13. Lugang Danpushe

According to *Caotun Town Records*, Yifu (Tainan Prefecture), Erlu (Lugang), and Sanmenga (Wanhua) were the three major ports in Taiwan in the past. Lugang was the largest cargo port in central Taiwan at that time. In the early days of the Japanese occupation when the material was scarce, the Taiwanese people divided the time for farming and transporting goods for the livelihood of each family, forming a walking route from Lugang to Puli Village. The porters carried goods on their shoulders from Lukang, Changhua, Kuaiguan, Shekou, Xianzhuang, Beitou, through Cao Xie Dun, and east to Puli. Cao Xie Dun is where they would rest, spend the night, and replace their straw sandals. It is a unique and real event in the "Lugang Danpushe", often mentioned in stories by Taiwanese people.

1.5 Object and Scope of the Book

1. Geographical scope

Caotun Town, Nantou County is slender from east to west and narrow from north to south. It faces Wufeng Township, Taichung County across Wuxi to the north, Guoxing Township to the east, adjacent to Nantou City and Zhongliao Township in the south, and connects Fenyuan Township, Changhua County to the west, shaped almost like a horizontal rectangle. There are no mountains in the territory, the climate is mild all year round, and there is no monsoon wind. Location: East longitude 120 degrees 37 min 45 s; 120 degrees 49 min 42 s. Latitude 23 degrees 56 min 48 s north; 24° 1 min 19 s. Area: 104.0327 square kilometers. Caotun Town used to be the domain of the Hoamya, Arikun, and Tausabata tribes. The large-scale cultivation of land by the Han people in the township began after the Qing Dynasty. The four family names of Hong, Li, Lin and Jian are dominant in the region. The direction of its settlement is roughly from west to east, all the way up the valley. In the early days of Japanese occupation, it was divided into Caoxiedun District, Xinzhuang District, and Tucheng District; amend to Caotun town after Taiwan Restoration in 1945. After Taiwan Restoration, Caotun Town remained located at the entrance and exit of Nantou County, with convenient transportation and rich products, suitable for permanent settlement, a large influx of residents from all ethnic groups formed a tightly knit community, which worked together to create an attractive living environment (Caotun

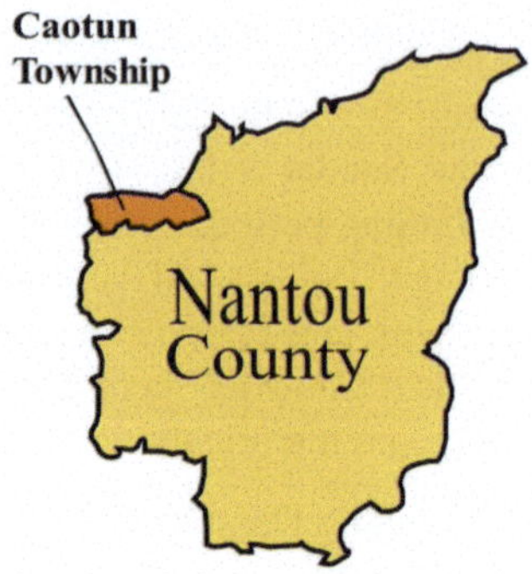

Fig. 1.1 Administrative map of Caotun

Township Office, Nantou County, 2022). According to household registration data, by the end of December 2001, there were 49,009 men and 48,352 women, with a total population of 97,064. The total number of households is 34,298, the total number of is 608 "Lin", and the whole town is composed of 27 "Li (Nantou County Government Civil Affairs Office, 2022)." At present, Caotun Town is the largest town in Taiwan. This book will limit the scope of research to the resources of Caotun Town's natural, production, human, cultural and landscape (Fig. 1.1).

2. The concept and form of the knowledge base of cultural and historical design symbols

Field investigations are carried out in this book with the purpose of constructing the knowledge base of cultural and historical design symbols based on the resources of natural, production, human, cultural and landscape, and inputting them into the knowledge base as examples. Due to the huge limitation of local resources, the most representative resources are selected first, and not all resources can be documented; the focus of the knowledge base is on the need to verify usability for both local and foreign designers.

3. Experimental samples

(1) For the experimental sample of the design and development mode of cultural and creative products, foreign and local designers were selected, the foreign designers were selected from students of the Department of Industrial Design of Donghai University, a total of 35 people, and the local designers were selected from the Cao Xie Dun local cultural and educational association, a total of 10 people. These groups mainly verify the feasibility and practicality of using the knowledge base for two groups of designers from different backgrounds.

(2) The experimental sample of cultural and creative commodities display mainly investigates whether foreign tourists and local residents identify local cultural and creative commodities as representative of the local culture and its related industries, as well as the willingness to purchase said commodities.

4. Cultural and creative products design

(1) In addition to verifying the design and development model of cultural and creative products with examples in this book, designers are also invited to the

design site to experiment with the students of the Development Department of the Cao Xie Dun Local Culture Education Association (local designers) and the students of the Department of Industrial Design of Donghai University (out-of-town designers).

(2) In reference to the materials used in the design of cultural and creative products, originally, the local cultural and creative products in Caotun were mainly straw. However, this book takes into account the many natural resources used in local industries in addition to straw. Therefore, this research uses ceramics as the main material, which is supplemented by straw, bamboo, rattan, and wood.

1.6 Subjects of the Book

1. Subjects of the First Questionnaire

The purpose of the first questionnaire survey is to verify the feasibility of using the knowledge base of cultural and historical design symbols by those with experience of related product development. The following two subjects are selected for in-depth investigation:

1. Students from the Department of Industrial Design of Tunghai University: Developers of cultural and creative products, using the "knowledge base of cultural and historical design symbols" as the basis for designs, in order to verify the feasibility of its application.
2. Cao Xie Dun Local Cultural and Educational Association: Select a specific local cultural and creative product development group, and understand the feasibility of the conceptual form of the cultural and historical design symbol knowledge base in the design of culture through a closed questionnaire survey.

2. Subjects of the second questionnaire

   The purpose of the second questionnaire survey is to plan the cultural and creative commodity exhibition, to verify the evaluation of foreign tourists and local residents, and to understand whether the product is a local cultural and creative commodity, which properly reflects the locality. For this survey, there are two groups of subjects:

   (1) Out-of-town tourists: An open-ended questionnaire was conducted for out-of-town tourists and residents to investigate the differences in their awareness of the local cultural and creative industries and commodities, as a reference for the construction of a knowledge base of cultural and historical design symbols.
   (2) Local residents: An open-ended questionnaire was conducted to explore and analyze the differences in the awareness of the exhibition content among local residents and long-term workers.

1.7 Main Research Methods

There are two main methods in this book, "Semiotics" and "scenario-oriented". Using the common methods of semiotic scholars to interpret symbols to establish the conceptual form of the knowledge base of cultural and historical design symbols and design and develop local cultural and creative products. Based on consumer evaluations, the design results of local cultural and creative products still cannot achieve a noticeable difference in the recognition of these products by the mass consumer; seeking the effect of the scenario-oriented design method, to strengthen the impression of the exhibition narrative, and effectively encourage the consumer's purchasing behavior and cultural awareness.

1.8 Framework and Description of the Book

This book is divided into four stages: Stage I involved identifying the problems concerning local goods, establishing a database prototype, and integrating the vast local resources; Stage II involved constructing a database of local resources based on the theory of semiotics, and providing a development model for local cultural goods, involved verifying the model by designing and developing goods ourselves, and conducted the first questionnaire survey; Stage III identified a difference in cultural cognition with cultural and creative products among consumers, and implemented the scenario-oriented design method in the planning of the cultural and creative product exhibition, while concurrently conducting the second questionnaire survey; and Stage IV involved drawing conclusions and offering suggestions. As explained in Fig. 1.2 below.

1. Phase 1: Through field investigation, it was found in academic writing that "cultural and creative products" caused difficulties in the development and promotion of the cultural industry, and it was learned that designers and consumers had insufficient understanding of the products. In the process of literature research, there was the goal of finding various research problems and proposing applicable methods to solve them.
2. Phase 2: Through semiotics and the theories of various scholars, conduct the analysis of relevant academic writings, construct the conceptual form of the local cultural and historical design symbol knowledge base, and design and develop local cultural and creative products. At the same time, conduct the first questionnaire to test the feasibility and practicality of using the knowledge base, the subjects of which include local and foreign designers.
3. Phase 3: In order to explore consumers' differences in cultural understanding as a result of cultural and creative products, the scenario-oriented design method will be adopted, and cultural and creative product exhibitions will be planned. At the same time, the second questionnaire survey will be carried out to test the

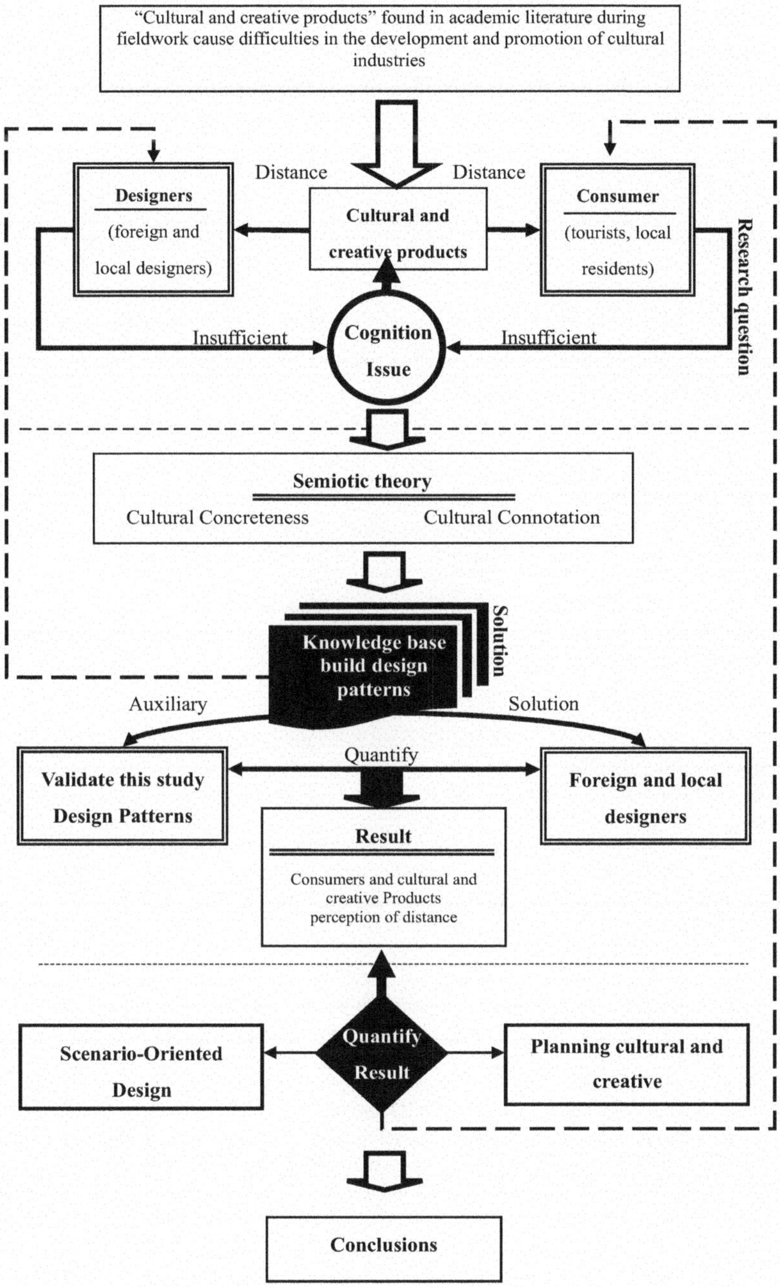

Fig. 1.2 Research process and core

preferences and awareness of the cultural and creative commodity exhibitions among foreign tourists and local residents.

4. Phase 4: Conclusions and suggestions, and describe areas for follow-up research and development.

References

Caotun Township Office, Nantou County. (2022). Geography. Retrieved form http://www.tsaotun.gov.tw/.

Council for Cultural Affairs, Executive Yuan. (2022). Culture industries. Retrieved form http://www.fjweb.fju.edu.tw/french-culture/cre.htm.

Chen, K.C. (2005). *The research of shop ecology and cultural goods of the Taiwanese museums.* Master's thesis. Tainan: Institute of Art Studies, National Cheng Kung University.

Chen, C. N. (1995a). *Overall community creation and cultural industry development.* Caotun Township: National Taiwan Craft Research Institute, pp. 4–9.

Chen, C. N. (1995b). *Culture, Industry Symposium and Overall Community Building China-Japan Exchange Exhibition Proceedings.* Taipei: Council for Cultural Affairs, Executive Yuan.

Department of Communication and the Arts (DCA). (1994). *Creative Nation: Commonwealth Cultural Policy.* Australia: Department of Communications and the Arts.

Executive Yuan. (2022). Local Creation. Retrieved form https://www.ey.gov.tw/Page/5A8A0CB5B41DA11E/9f7f9ad7-17e4-482a-9956-a1534d8a4ad1.

Gao, X. A. (2002). *Sociology of pop culture.* Taipei: Yang Zhi Culture.

Lin, H. C. (2011). *A comparative study on the governance models of creative and cultural industry centers in Taiwan.* Master's thesis. Nantou County: Department of Public Policy and Administration, National Chi Nan University.

Ming, L. G. (2000). *Reconstruction of local cultural industry and resource integration.* Seminar on experience exchange of local cultural industry revitalization program and achievement album of talent training program. Taipei: Council for Cultural Affairs, Executive Yuan.

Ministry of Culture. (2008). The third phase of community development and village cultural development plan. Retrieved form: https://www.moc.gov.tw/content_268.html.

National Culture and Arts Foundation. (2004). *A book on the practice of cultural and creative industries.* Taipie: Business Weekly.

Nantou County Government Civil Affairs Office. (2022). Demographic information management platform. Retrieved form https://household2.nantou.gov.tw/00home/index.php.

Neuman, L. W. (2005). *Social research methods: Qualitative and quantitative approaches.* Prentice Hall.

National Development Council. (2022). Design flip, local creation. Retrieved form https://www.ndc.gov.tw/Content_List.aspx?n=4A000EF83D724A25.

Su, Z. Y. (2009). Community empowerment/development, Encyclopedia of Taiwan, Ministry of Culture. Retrieved form http://nrch.culture.tw/twpedia.aspx?id=3972.

Wang, L. J. (2003). *Economic discourse in cultural politics: From elite culture to mass economy?* (pp. 1–18). Annual Meeting of the Society for Cultural Studies, By Culture, Academic Symposium.

Wang, L. J. (2005). Economic discourse in cultural policy: From elite culture to cultural economy. *A Journal of Cultural Studies, 1*, 169–195. https://doi.org/10.6752/JCS.200509_(1).0007

Weng, X. D., & Miyazaki, K. (1996). *The flower of the heart—Concepts and examples of the overall building of Japanese communities*, Caotun Township: National Taiwan Craft Research Institute.

Chapter 2
Research Methods

The research content is roughly divided into two sections: qualitative and quantitative. The qualitative research method section includes field research to identify the problems of local designers in the development and promotion of products. While the semiotic analysis method is implemented to analyze common theories used by various scholars to interpret symbols, and to build a knowledge base of cultural and historical symbols to assist in the design and development of cultural and creative products. The first questionnaire aims to analyze the feasibility of the knowledge base, but the lack of cultural understanding among consumers regarding cultural and creative products remains. Thus, the scenario-oriented design method is used to plan the cultural and creative product exhibition, to be used as the basis for the second questionnaire, which analyzes the cultural understanding of consumers regarding each locality within the exhibit. Definitions and explanations of other research methods, including the participatory observational method, community study method, method of data collection, are as follows (Fig. 2.1).

.

2.1 Research Methods and Process

See Fig. 2.1.

2.2 Qualitative Research

Qualitative research refers to constructing the cultural meaning of reality in society, it is defined by its focus on the process of interactive events, the importance of authenticity, its ubiquitous and obvious value, its limitations due to context and few case subjects, thematic analysis, and heavy involvement from the researcher

M.-F. Wang, *Creative Product Design With Cultural Codes*,
https://doi.org/10.1007/978-981-19-6767-2_2

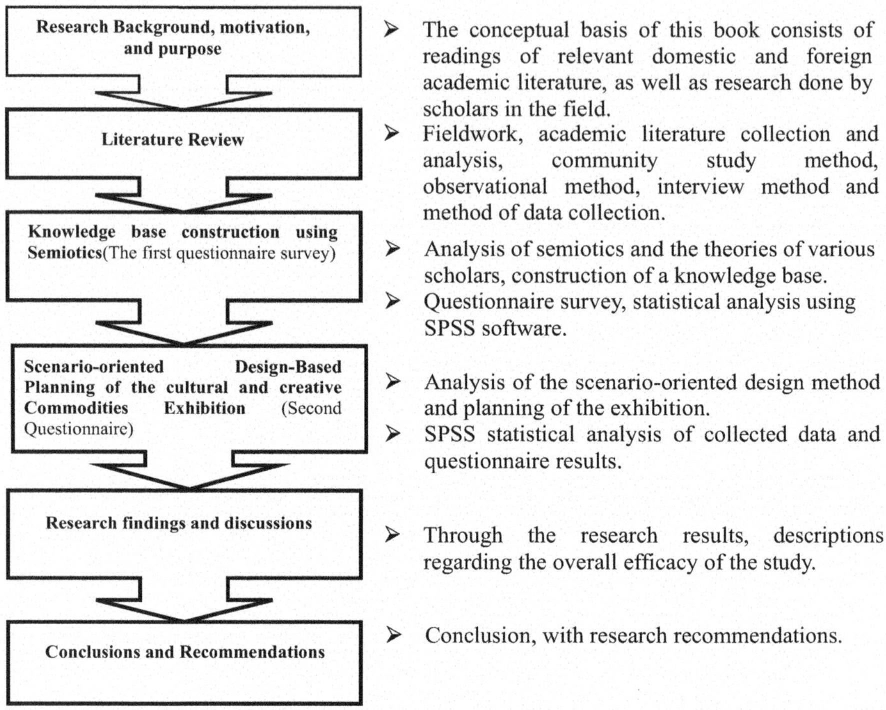

Fig. 2.1 Research methods and steps

(Neuman, 2005). Qualitative research is a category name relative to "quantitative research", not a single or specific research method. Qualitative research is suitable for identifying and explaining some little-known aspects below the surface of a problem, or taking a new and creative approach to some well-known issues. At the same time, it can also provide details through the in-depth exploration of complex situations that quantitative research cannot describe. Using the inductive analysis method of qualitative research, one can ascertain the meaning and context of various situations, and make interspersed connections between oral interviews, academic literature, and the entire qualitative process. Ultimately, the process allows the researcher to trace information layer by layer to establish a context, which will enable the determination of the most complete and accurate result (Liu, 1996; Liu, 2002). The characteristics of qualitative research are as follows:

(1) Seeing the world through the eyes of the subject: What the researcher wants to interpret is the social facts and reality constructed by the subject, not the social facts and reality as understood by the researcher. The researcher must interact with the subject for a long time to develop the same understanding in order to fully comprehend the thinking of the subject.
(2) In-depth description: Qualitative research must make detailed records of the subjects, setting, and context, so that readers can correctly understand the

meaning of the contextual relationship which results from the interaction between researcher and subject.

(3) Contextualism: Qualitative research emphasizes that the meaning of the result is not fragmentary and isolated, that the result has its proper context, and must be viewed from a holistic perspective.

(4) Process: Qualitative research believes that society is constantly changing, and people's lives are composed of a series of interrelated activities. Qualitative research assigns great importance to the process of change in people's daily lives.

(5) Flexibility: Qualitative researchers do not have a pre-determined stance in order to avoid preconceived ideas, and mostly use open or unstructured methods in a cyclical manner to collect and analyze data. The key to qualitative research lies in the researcher's ability to empathize with testimonies. In qualitative research, interpreting subject responses during the process of understanding a subject is not a single-sided interaction of stimulus–response, but a two-way exchange of continuous reaction and response incorporation.

Qualitative research is the combined fusion of dialogue from many holistic experiences or horizons. Fairclough pointed out that there are three levels of discourse: text, interaction and context. Therefore, discourse analysis also has three levels of work: description, interpretation, and explanation. Description is the analysis of the formal nature of the text, interpretation explores the relationship between the text and interpersonal interactions, and explanation studies the relationship between interpersonal interaction and the social context (Fairclough, 1983, 1985, 1989, 1992, 1993, 1995; Fairclough and Wodak, 1997). This research uses related qualitative research methods such as field investigation, the participant observation method, the participant interview method, semiotic theory, the scenario-oriented design method, the community study method, and various data collection methods.

1. Field Survey

Robert M. Emerson believes that the field research method is the most important data collection strategy in the qualitative research method. Through researchers inhabiting the world of others, studying the natural patterns of behavior of people in their daily lives, learning directly about how they live, how they speak and act, and why they feel confused and distressed. Moreover, a field researcher must be able to empathize and see the world from their subject's point of view, in order to understand the meaning of their perspective (Chava & David, 1981, 1999). In a broad sense, all studies conducted in the field can be called "field research", but in a narrow sense, field work refers specifically to archaeological discoveries and ethnographic investigations in the field of anthropological research (Ye & Ye, 2011). Earl Babbie points out that a key characteristic of field research is that "it provides the researcher with a degree of understanding of the research point of view (Earl, 2001). By observing and learning a social phenomenon as completely and directly as possible, it can also be understood more deeply and comprehensively". Anthropologists were the first to use field research as their research method. Later, sociologists adopted the research method and also referred to the area of their research as the “field”. Field research (surveying) requires the collecting of data and the conducting of research in a specific academic field over an extended period of time.

Why is fieldwork considered qualitative research? The purpose of field research is to understand the subject in a natural context, or to explore an individual’s psychology, consciousness, and behavioral processes regarding a particular subject, towards which their attitudes and behaviors often cannot be quantitatively measured. The biggest difference between qualitative research and quantitative research in form is that qualitative research rarely sets and manipulates variables, but instead uses interviews, observations, and records of field events (Chen, 1990; Stewart, 1998).

So what kind of research topic is suitable for qualitative research? For example, if you want to explore the interaction between local cultural festivals and tourists, or further understand the history of Caotun and the significance of cultural and creative products, or other similar phenomena which cannot be investigated through quantitative methods, it is easier to carry out research with qualitative methods instead. This research uses the field study method to collect “industry, nature, humanities, culture, landscape” to collect and classify data regarding the five resources within a locality, and ultimately transform the data into a knowledge base of cultural and historical design symbols (Table 2.1).

2. Participant observation

Participant Observation is one of the main methods of collecting data in qualitative research, through which the observer (researcher) and the subject live and work together, with the observer listening and watching the subject’s words and actions via the direct experience of close mutual contact. The situation under which this observation takes place is relatively natural, and the observer can not only gain a more specific understanding of the local social and cultural phenomenon through the subject’s perspective, but also can probe deep into the culture of the subject and understand their interpretation of the meaning behind their behavior. The definition of participant observation is field observation or direct observation, with the goal of achieving a scientific understanding of a group. The researcher (or investigator)

Table 2.1 Five major local resources

Production resources	Natural resources	Human resources	Landscape resources	Cultural resources
Straw Sandals and other investigations	Jiujiu Peak and other investigations	Caotun Township Office, Nantou County, National Taiwan Craft Research Institute, the Cao Xie Dun Local Cultural and Educational Association, etc.	Furong Hall and Eight-diagrams Kiln	Annual international straw craft culture festival

establishes and maintains a multi-faceted and long-term relationship within that group to facilitate the process of their research. Any study which needs to understand the context of processes, events, relationships, and social environments would adopt participant observation (Hu, 2008).

The reason for using the participant observation method in this study is to observe the communication and learning of each design team. In the process, the behavior of "people" and related influences are used as the research content, and the observed evidence is directly described and recorded, which is an important source of research data. The most authentic first-hand information can only be obtained by heavily investing oneself in the observation process, and researchers need only interpret the results and deconstruct the context under which they were collected (Wang and Wang, 1998). The participant observation method was used in the examination of local product design and development teams, for example: observing the Cao Xie Dun local cultural and educational association, their understanding and application of the local cultural and historical design symbol knowledge base. As well as using the development model of cultural and creative products to explore the core problems faced during the development of local cultural and creative products (Table 2.2).

3. Participant interviews

The interview method is an activity in which the researcher seeks out, interviews, holds conversations with, and asks various questions towards a subject. The interview is a research-based conversation, a research method in which researchers collect first-hand information from the subject through oral conversation. In terms of the degree of control the researcher has over the subject, interviews can be divided into three types: structured, unstructured, and semi-structured (Chen, 2002). An interview is another important way of collecting data in this study, as a type of targeted research dialogue. During an interview, the researcher and the subject engage together in the process

Table 2.2 Subjects

Participant observation	Subject of observation	
	Tunghai University Students (Foreign Designers)	The Cao Xie Dun Local Cultural and Educational Association (Local designers)
Participant observation	Students from the University Departmentof Tunghai University. Develop cultural and creative products using the knowledge base, and observe how designers from other places understand the local culture of Caotun, assisting designers in their use of the knowledge base	First, establish a relationship with the Cao Xe Dun Local Culture and Education Association, and understand the application of the knowledge base of local cultural and historical design symbols. At the same time, through the cultural and creative product development model, identify core problems faced by local designers when developing local cultural and creative products

of "constructing" the social scene. Through collecting first-hand information about subjects through the methods of continuous interviews and relaxed discussions, we are able to quickly understand the feasibility of the application of the knowledge base as well as how the product development model operates within a given design team. The interview is a survey method in which the researcher conducts data collection through engaging in direct contact with the subject. It can collect more in-depth data, especially through face-to-face interviews, where the researcher can observe the reaction of the subject in real time, ask follow up questions, and obtain more specific information or leads (Chen, 2002). In addition, through interviews with leaders of local cultural organizations, historians, local elders and local residents, this study obtained oral information such as the origin of local place names, along with numerous other stories and legends to assist the writing of background information using the scenario-oriented design method.

4. Semiology

Semiotics is a method of linguistic analysis which is a theory about symbols or symbolic processes (Wang and Wang, 1998). The concept of semiology was first proposed by the Swiss scholar Ferdinand de Saussure, who established a theory of symbols that he called "Semiology". Among them, he regarded a symbol as an entity with two faces, including "Signifier" and "Signified" (Lin, 2000). For example, the shape, color, and symbolic culture of Caotun cultural and creative products can be referred to as the "signifier" of a symbol; and the meaning, symbolism, and connotation that these elements truly wish to express are called the "signified" of a symbol. Saussure holds that the character of the symbol is by no means entirely arbitrary; it is not meaningless, because there is a natural connection between the "Signifier" and the "Signified" (Bevger, 2004). Synthesizing the various semiotic theories and methods of Charies Peirce, Ferdinand de Saussure, Charies Morris, Roland Barthes, Umberto Eco and Katsumi Hoshino, analyzing the formation of cultural and creative goods, and the process through which culture is transformed into a product. Cultural symbols will be selected, shaped, translated, conveyed and integrated into various products (Wang & Cheng, 2021). Therefore, this book will establish the conceptual form of the knowledge base of cultural and historical design symbols, and promote the conversion of local cultural symbols into products, and thus use the aforementioned method for the purpose of this study.

5. Scenario-oriented design

It has been 10 years since Acer introduced the scenario-oriented design method from IDEO in 1990. The advantage of "Scenario-Oriented Design" is that the planner (PM), the designer (ID) and the engineer (R/D) can all communicate through a script which, through each party's understanding of different situations, aids in understanding and identifying the various needs of potential customers. In this way, this approach is appealing in both a rational and emotional sense (Yu et al., 2000). Artificial intelligence scientist Roger C. Shank explained in his book *Tell Me A Story* that human memory stores thousands of scripts, these scripts are the accumulated wisdom of our experiences in dealing with life events. When faced with particular circumstances or environments in life, we retrieve the corresponding scripts in response. If we combine these experiences with the corresponding scripts we use, they become stories. Shank believes that the use of scripts is a kind of wisdom human beings use to solve problems in daily life. Using scripts reduces the burden of thinking, and using correct scripts reduces risks and improves adaptability in various situations. If solving complex mathematical problems is an example of human intelligence, making good use of scripts would be an example of human wisdom (Yu et al., 2000). This study also uses the method of scenario-oriented design to write characters, maps, and situational scripts to plan the purpose of cultural and creative commodity exhibitions.

6. Community study method

A method of exploring community life, value systems, structures, functions, problems, and changes from the relationships among community constituents. The community study method was first adopted by cultural anthropologists, and later frequently used by sociologists and social psychologists. Community refers to a relatively independent social area composed of people and collectives who occupy a certain area. Community research methods usually use data collection, the participant observation method, in-depth interviews, the sociometric method, family tree collection, the case study method and content analysis during research (Wang and Wang, 1998). This study will adopt this approach to explore the role of the Cao Xie Dun local cultural and educational association, as well as organizational members and local designers of the Caotun community in various design activities.

7. Data collection method

This book collects and analyzes relevant academic literature and data regarding the research topic. The data collection method is one of the simplest exploratory research methods, which is constructed by collecting research conducted by other scholars, questionnaires, academic literature, relevant media clippings and seminars held by various organizers, and extracting relevant research topics to be the basis of this study through analyzing the research results and follow-up suggestions (Yang et al., 1989).

2.3 Quantitative Analysis

Quantitative analysis is used to measure objective facts, it focuses on variables, reliability is essential, neutrality is of paramount importance, studies are not affected by context, and contain multiple case subjects, results are analyzed with statistical analysis, and the researcher maintains a neutral stance. Most quantitative techniques are agglomeration of data, they aggregate data to present the whole picture of a given phenomenon. Quantitative research techniques share the language and logic of positivism, which sets them apart from research techniques based on other research orientations. The language of quantitative research is a language of variables, causality, and hypotheses. Quantitative research designs use deductive reasoning. When using quantitative research, the final step is to test the hypothesis against empirical evidence (Neuman, 2005). Quantitative research methods used in this study include the questionnaire survey method, statistical analysis and collation using the SPSS software, as well as other methods.

1. Questionnaire survey method

The researcher will create a questionnaire based on the matters he seeks to study, and distribute or mail it to the respondents (subjects) to answer. This research method of using questionnaires as a data collection tool is called Questionnaire Survey. The two questionnaire survey methods include mail questionnaires and face-to-face questionnaires. The "advantages" of questionnaire surveys include (1) when surveying a large region, questionnaires are economical and convenient, and a large quantity of sample data can be obtained. (2) Questionnaire subjects are unrestricted and anonymous, which makes questionnaires suitable for inquiring into sensitive issues. (3) The chance of bias error is low. (4) Respondents have enough time to think and provide a thoughtful answer. (5) Questionnaires have a low cost. Conversely, the "disadvantages" of the questionnaire include: (1) The questionnaire response rate can be low. (2) Questionnaire questions must be simple (Wang, 2002).

SPSS statistical software is a type of software that provides data collation and service solutions, to obtain the most valuable information from questionnaire data. Combined with the statistical functions of SPSS, the questionnaire answers can be converted into meaningful information, and the opinions of the questionnaire respondents can be better understood. The one-way analysis of variance test procedure was used to further select an appropriate method to conduct a post-hoc comparison, which was performed to determine which pairs of treatment levels the samples had significant differences in mean. Questionnaires can be divided into three types: structured, unstructured, and quasi-structured (Wang, 2002). Through utilizing the scenario-oriented design cultural and creative product exhibition to display the various designs, exhibition layouts, and products that express local culture, stories, history, symbols, and symbolic content, this book hopes to increase foreign tourists' and local residents' understanding of local cultural and creative product exhibitions. Detailed statistical data has been collected by running selected survey data through the SPSS statistical software. At the same time, a closed questionnaire targeted at both foreign and local designers is used to investigate the feasibility of developing cultural and creative products using the knowledge base (Table 2.3). Another open-ended questionnaire, such as a survey on the awareness of cultural and creative commodity exhibitions, is based on the responses of local residents and foreign tourists (Table 2.4).

2. Statistics Package for the Social Sciences (SPSS)

The full name of SPSS is Statistics Package for the Social Sciences. It is a software program developed by Nie Hull Jenkins and Steinbrenner of the University of Chicago as well as Bent of the University of Alberta for the purpose of data processing. It can perform descriptive univariate statistics, multivariate statistics, nonparametric statistics, etc. Using SPSS for statistical analysis is mainstream

Table 2.3 Respondents to questionnaires with Close-ended questions

Questionnaire subject	Professional field	Questionnaire type
Sophomores of the Department of Industrial Design at Tunghai University (Foreign Designers)	Industrial product design, human factors and ergonomics design, market analysis, material application and research, computer 3D rendering and modeling technology	A structured questionnaire survey conducted among select individuals with practical experience in product development
The Cao Xie Dun Local Cultural and Educational Association (Local Designers)	Technicians at the National Taiwan Craft Research and Development Institute, Community Cultural Guides, Artisans Department within the Grassroots Alliance, local dialects, Practical cultural and creative Product Development within the Grassroots Alliance, archaeological studies, education of straw patchwork dyeing and weaving, community building, financial management, field studies	

Table 2.4 Respondents to questionnaires with open-ended questions

Questionnaire location and context	Nature and space of the exhibition venue	Questionnaire subject	Questionnaire type
Cultural Affairs Bureau Taichung City Wen Ying Hall	Pure exhibition viewing	Tourists locals	Unrestricted number and type of subjects, unstructured questionnaire type, survey on understanding of cultural and creative products
Caotun Youth Activity Center (Dao Cao Local Culture Center)	Cultural and creative product sales area Pure exhibition viewing DIY creation of cultural and creative products	Tourists locals	
Caotun Township Library	Pure exhibition viewing	Tourists locals	

among larger computing platforms, especially in research and academic institutions. Although the PC version does not fully support the original functions of SPSS/PC+, SPSS remains the most effective tool for statistical analysis when combined with the PC platform, and the software is still able to process a high quantity of data when operating in the PC environment. Furthermore, the interactive operating system and execution thread count results in a piece of software which has a low disk space requirement (Dodge, 2008; George & Mallery, 2010). The study detailed in this book will use SPSS to carry out research work such as cross analysis, narrative analysis and factor analysis.

3. Data collation method

Methods for scientific classification and aggregation of raw data obtained from statistical surveys will be chosen according to the original purpose of the data collation. Whether the results of data collation are correct or not directly affects the accuracy of the overall quantity description and the authenticity of the analysis to follow. There are two forms of organization and implementation of data collation: (1) Level-by-level sorting, which refers to sorting survey data from bottom to top according to a given statistical management system. (2) Centralized sorting, that is, all the survey data are centralized that organizes the statistical survey for sorting using highest-level processing (Wang and Wang, 1998). In practical research, these two forms can also be used in combination with one another, allowing for the most basic statistical indicators to be unified and aggregated. Through the level-by-level and centralized sorting methods, the statistical data obtained in this study are able to have an accurate quantitative description and authentic, verified results.

To summarize, this book is divided into a qualitative research section and a quantitative research section. The two research methods have their own strengths and complement each other to achieve the effect of truncating the resulting data from the study. At the same time, these two research methods aid in clarifying the definition of relevant terms, apply appropriately in potential follow-up research, and achieve a practical application of research methods.

References

Bevger, A. A. (2004). *Shop 'til you drop: Consumer behavior and American culture*. Rowman & Littlefield Publishers.
Chen. X. M. (2002). *Quality research in social sciences*. Taipei: Wunan.
Chen, B. Z. (1990). *A new orientation in educational research methods—Qualitative research methods*. Taipei: Nan Hong.
Chava, F. N., & David, N. (1981). *Research methods in the social sciences* (2nd ed.). St. Martin's Press.
Chava, F. N., & David, N. (1999). *Research methods in the social sciences*, Enskede: TPB.
Dodge, Y. (2008). *The concise encyclopaedia of statistics*. Springer-Verlag.
Earl, B. (2001). *The practice of social research* (9th ed.). Wadsworth Thomson Learning.
Fairclough, N. (1983). *Language and power*. Longman.
Fairclough, N. (1985). Critical and descriptive goals in discourse analysis. *Journal of Pragmatics, 9*, 739–763.
Fairclough, N. (1989). *Language and power*. Longman.
Fairclough, N. (1992). *Discourse and social change*. Polity Press.
Fairclough, N. (1993). Crtical discourse analysis and the marketization of public discourse: The universities. *Discourse and Society, 4*(2), 133–168.
Fairclough, N. (1995). *Critical discourse analysis: The critical study of language*. Longman.
Fairclough, N., & Wodak, R. (1997). *Critical discourse analysis, in T. van Dijk (ed.)*, Discourse as Social Interaction. 2, 258–84. London: Sage.

George, D., & Mallery, P. (2010). SPSS for Windows step by step: A sample guide and reference 17.0 update. Boston: Allyn & Bacon.

Hu, Y. H. (2008). *Qualitative research: Theories, methods and examples of local women's research.* Mighty Current Books Company.

Liu, Z. D. (1996). *Ethnographic research methods and examples* (pp. 173–193). Taipei: Mighty Current Books Company.

Liu, Z. D. (2002). *The controversy of quantitative and quality social research and the future trend and outlet of social research.* Mighty Current Books Company.

Lin, M. H. (2000). Sign and Code in Product Design. *Journal of Design, 5*(2), 73–81.

Neuman, L. W. (2005). *Social research methods: Qualitative and quantitative approaches.* Prentice Hall.

Stewart, E. (1998). *The Ethnographr's method*. Sage.

Yang, G.S., Wen, C.Y., Li, Y.Y., & Wu, C. X. (1989). *Basic concepts of scientific research, research methods of social and behavioral sciences* (Vol. 1, 13rd ed.) Taipei: Tung Hua Book Co., Ltd.

Ye, Z. C., & Ye, L, C. (2011). *Research methods and paper writing.* Taipei: Shangding Publishing House.

Yu, D. Z., Lin, W. Q., & Wang, J. Q. (2000). *Scenario-oriented design: New law of product and service design in the information age.* Garden City Publishing House.

Wang, H. H., & Wang, X. K. (1998). *Encyclopedia of scientific methods.* Taipei: Enkai.

Wang, M. F., & Cheng, C. F. (2021). Hakka Faith Sanshan king cultural and creative product design and research. In *National science forum, Pingtung: National museum of marine biology and aquarium* (pp. 229–232).

Wang, B. J. (2002). *SPSS for windows and behavioral science research.* Taipei: Psychology.

Chapter 3
Literature Review

The development and design of cultural and creative products in local industries in Taiwan is still not yet fully matured. Based on this situation, this study summarized the categories of cultural and creative products and definitions to explain the spiritual meaning and connotative meaning of cultural and creative products from their original definitions. From the categories of cultural and creative products all over Taiwan, we explored the cultural elements displayed in the Museums, ethnic groups, the local industries, and classifies the three connotations of local cultural and creative products, namely, symbolism, historical narrativeness, and local uniqueness, to entail the research with the initial design and development model for the local cultural and creative products.

3.1 Field Survey

The purpose of the field investigation is to understand the current situation of the local cultural industry in Caotun, and to obtain a deep understanding of local cultural organizations and local residents, to document, investigate, narrate, interview and record the issues related to the cultural industry as they occur in the local culture. The following is a classification survey of Caotun's local resources, natural, production, human, cultural and landscape.

1. Classification of natural, production, human, cultural and landscape resources

Caotun has rich local cultural and industrial resources. The most distinctive local cultural and creative industries can be explained by properly classifying and planning the resources. According to the book *Theory and Examples of Overall Community Creation II*, planned and edited by the Cultural Construction Committee of the Executive Yuan (Ministry of Culture of Taiwan in the modern day) in 1999, based on the original text produced by the National Taiwan Craft Research and Development Institute in conjunction with the Miyazaki Laboratory at Chiba University in Japan,

M.-F. Wang, *Creative Product Design With Cultural Codes*,
https://doi.org/10.1007/978-981-19-6767-2_3

the book pointed out: "There are many local resources, and the overall construction of the community starts from a new understanding of local resources. Among them, the local traditional crafts are especially irreplaceable. Local traditional crafts are a local living industry, and are now receiving extensive attention." To organize the local knowledge base as a resource, it is necessary to link local industries which rely on the five major resources of "land, scenery, people, culture, and production". Here, "Land" refers to natural resources; "Scenery" refers to landscape resources; "People" refers to human resources; "Culture" refers to cultural resources, and "Production" refers to production resources. natural, production, human, cultural and landscape are not five independent resources, but are all interconnected. Among them, "Culture" is the soul of various resources: the production of local "Culture" is the process in which local people use the resources of "land" to create "landscape" and "production". These aforementioned processes are the development of local history and culture, local life culture, and local customs and religious beliefs (Council for Cultural Affairs, Executive Yuan, 1999; Huang, 2001). Therefore, the generation of "culture" is a process in which the local "people" "land", "scenery", and "production" are used, intertwined, and produced. The scope of the fieldwork in this research is precisely the industries that connect the five local resources of "land, scenery, people, culture, and production", and define how they are interrelated rather than independent resources. The following is a correlation diagram, which displays the relationship between land, scenery, people, culture, and production (Fig. 3.1).

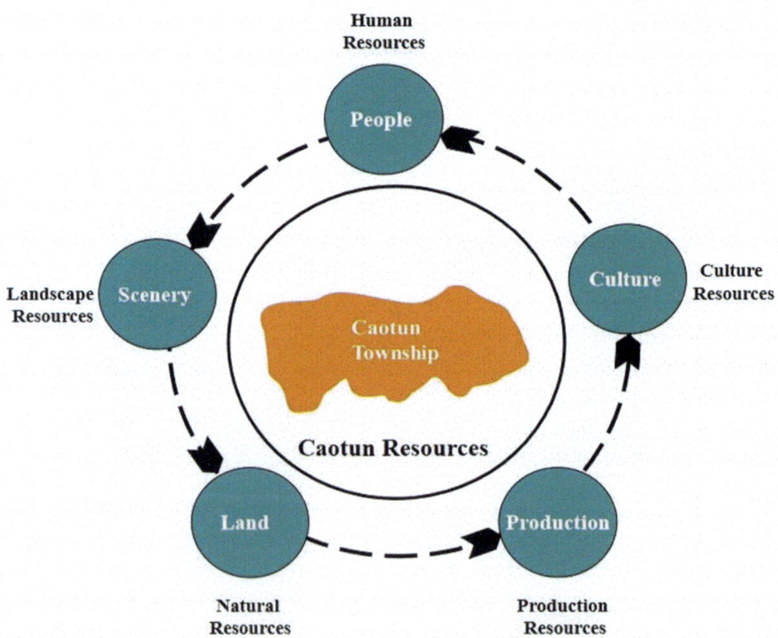

Fig. 3.1 Correlation map of natural, production, human, cultural and landscape in Caotun

Table 3.1 Field survey projects of the five resources: natural, production, human, cultural and landscape will include detailed site records and photographs on the history of natural resources such as Caotun Jiujiu Peak Nature Reserve, Qidai Mountain, Dahu Mountain, Tombstone Mountain and Dazhangshugong, etc. The resource characteristics and historical documents on landscape resources include the famous scenic spots in Caotun, Taiwan Craft Research and Development Institute, Furong Hall, Dun Lun Hall, Dengying Academy, Eight-diagrams Kiln, Shuangshi Suspension Bridge, Eighteen Dragon Pillars, Straw Craft Culture Museum, related crafts and artist studios, etc. As for human resources, the survey will include documents related to as well as interviews with famous local historical figures, administrative staff of Nantou County Caotun Township Office, Taiwan Craft Research and Development Institute, members of Cao Xie Dun Local Culture and Education Association, students of cultivating craft inheritance in the Straw Craft Culture Center, historians, heads of cultural and historical studios, local senior citizens, community builders, craftsmen, artists, Caotun tourists and local residents. In the section of cultural resources, researchers will participate in and observe international straw culture festivals, and investigate the formation of Caotun place names, the allusions of Lugang Danpushe Society, local legends and local religious and cultural festivals, etc. Finally, local production resources are explored in the range of straw sandals, lychees, betel nut, grapes and rice. Although the surveys conducted in this book are limited, they are still able to highlight some of the problems faced by Caotun, to serve as an extended basis for future research.

2. Caotun SWOT analysis

Caotun, a town which used to be defined by its traditional industry, despite retaining related institutions such as the Cao Xie Dun Local Culture and Education Association, has been making an effort to follow programs regarding the transformation of craft towns, as well as the concept of "industrial culture, cultural industrialization", initiatives which have been vigorously promoted by the Taiwan Crafts Research and Development Institute along with the Nantou County Caotun Township Office over the past 20 years. However, at present, there are still many problems and challenges to be learned from and tackled in the straw craft industry. The following section will summarize and analyze the contents of field surveys, verbatim transcripts of interviews, and academic literature pertaining to Caotun, and analyze the advantages, disadvantages, opportunities, and threats of the future development of the straw culture industry in Caotun. Finally, the section will propose research questions with the goal of solving the aforementioned issues (Table 3.2).

In the future, Caotun will focus heavily on the development of its straw culture industry. For this development to take place, Caotun must tackle the three main problems, illustrated in the table above, which its straw culture industry currently faces: festival, design, and product issues.

Table 3.1 Field survey on natural, production, human, cultural and landscape in Caotun

Five resources	Survey item
Natural resources	Detailed on-site records and photos of natural resources such as Jiujiu Peak Nature Reserve, Qidai Mountain, Dahu Mountain, Tombstone Mountain, Da Zhangshu Gong, etc., and the historical stories of how these resources were developed
Landscape resources	Includes famous scenic spots in Caotun, Taiwan Craft Research and Development Institute, Furong Hall, Stew Lun Hall, Dengying Academy, Eight-diagrams Kiln, Shuangshi Suspension Bridge, Eighteen Dragon Pillars, Straw Craft Culture Museum, related crafts and artist studios and other landscape resources, characteristics and historical documents and records
Human resources	Famous local historical figures
	Local Caotun Town Office government officials' policies on craft town development
	The promotion policy of the administrative staff at the National Taiwan Craft Research and Development Institute
	The methods used by the Cao Xie Dun local cultural and educational association and the Culture Management Organization
	The Straw Craft Culture Museum, an overview of the consumption of cultural and creative products and the inheritance of artisans cultivated by various workshops
	Historians, heads of cultural and historical studios, local elders, community developers, craftsman, artists, Caotun tourists and local residents
Culture resources	Survey the international straw culture festival, the formation of place names in Caotun, the allusions of Lugang Danpushe Society, local legends and religious and cultural festivals
Production resources	A survey of the areas around the production of products such as straw sandals, lychees, betel nut, grapes and rice

1. Festival issues: Local cultural short-term celebratory festivals cannot promote long-term cultural business opportunities.
2. Design issues: Most of the local residents serve as product designers. Local residents have a high understanding of the local culture, but lack the design expertise necessary to enhance the value of products.
3. Product issues: Straw commodities generally cannot have the qualities of being historic, storied, and unique. Through developing products in a diversified direction by simultaneously combining function with the local cultural background, and improving the overall sophistication of the final product, there will be a positive effect on the promotion and marketing of subsequent products.

Industry, government, and academia have invested a lot of resources to assist the traditional handicrafts of Caotun, which has been considered a "sunset industry" in the past few decades. However, only looking for short-term economic benefits in celebratory festivals cannot make up for the competitiveness the industry has lost in

Table 3.2 SWOT analysis on future development of the straw art industry in Caotun (Chen, 2005; Wang, 2007)

Strength	Weakness
1. Possesses a rich collection of production, natural, humanistic, cultural, and landscape (scenic) resources 2. A unique historical place name in Cao Xie Dun 3. Basic seed teacher teams have been established 4. The recognition and consensus among local residents regarding straw culture is gradually on the rise 5. The Straw Culture Festival has gradually grown to become an important annual celebration in Caotun township 6. The straw industry is rich in resources	1. There is a lack of planning regarding the overall image of straw products 2. Cannot be mass-produced with mechanization 3. Material processing and handling is difficult 4. The difference between straw culture and the culture of other traditional industries is not noticeable 5. Lack of a local learning environment 6. Short-term, celebratory festivals are unable to promote long-term cultural business opportunities 7. Straw products generally lack local cultural connotations 8. Straw products are not highly delicate or well crafted 9. The functions of straw products are similar to those which already exist on the market, and generally have no outstanding features
Opportunity	**Threat**
1. The input of production, government, and academic resources 2. Future prospects of tourism and cultural and creative industries appear positive 3. The rise of local cultural awareness 4. Leisure time among Taiwanese people is increasing 5. The concepts of nature and environmental protection have been accepted by a large portion of the populace	1. Straw products are not easy to preserve 2. The development of straw products is limited by the type of products which can be produced 3. The process of passing on straw culture is not easy 4. Lack of full-time research institutions 5. Lack of effective marketing channels 6. There is a lack of a mechanism for reinvesting profits in the locality 7. Lack of professional design developers who can enhance product value

the market over time in the long-run. Therefore, this study considers the possibility of enhancing competitiveness through cultural and creative product design. Taiwan's cultural industry can be quantitatively defined and this industry is what the "sunset industry" referred to. Cultural and creative products must be superior to alternate products in terms of practicality, quantity and consumability, otherwise they will inevitably be eliminated from the market. At the same time, through field investigation, we aim to understand the issues plaguing local industries in Caotun and solve the design dilemma of local cultural and creative products. The following section will collect and analyze academic literature with topics focusing on an overview of relevant cultural industries at home and abroad.

3.2 Domestic and International Development Trends of Cultural and Creative Industries

1. Characteristics and development trends of foreign cultural and creative industries

According to the United National Educational, Scientific and Cultural Organization (UNESCO), the "cultural and creative industries" are divided into three categories: cultural products, cultural services and intellectual property rights. "Cultural products" refers to books, magazines, multimedia products, software, records, films, videotapes, sound and light entertainment, craft and fashion designs. "Cultural services" include performance services (theaters, opera houses and circuses), publishing, publications, newspapers, communication and architectural services, as well as audiovisual services (movie marketing, television, radio programs and home video, all aspects of its production such as reproduction and photocopying, film exhibitions, even the ownership and operation of cable, satellite and broadcasting facilities or cinemas) and library services, archives, museums and other services (United National Educational, Scientific and Cultural Organization, 2014).

Today's cultural and creative industries take cultural and creative products as the starting point, and take the origin of "creativity" as the concept, and then extend outward to combine with other industries, creating a trend where the products produced are more and more extensive. David Throsby's description of the cultural industry model is as follows: Traditional arts are the core of this industry model: music, dance, drama, literature, visual arts, handicrafts, and new types such as video art, performance art, computer and multimedia art, etc. Each art form can be an industry of its own, and we often see them that way, but usually not just in reference to the original creator. For example, when we say "music industry", we refer to various players, including composers, performers, publishers, record companies, distributors, marketers, retailers, fan clubs, etc. yet the core of the industry still revolves around musicians who create the original product. "Cultural and creative industries" take "creativity" as the concept and design as the method, which is far from the traditional "cultural industry" (Throsby, 2001). "Cultural industry" refers to the use of modes of non-referential production through various business practices to achieve the "standardization" and "Pseudo-Individualization" of cultural entities. The use of these modes of production is to achieve maximum economic efficiency rather than the realization of cultural ideas and artistic promotion. Therefore, in the aggregate market for producers, the cultural industry uses clever techniques through the usage of advertising and entertainment industries as intermediaries to "unify" the needs of consumers. Unified demand creates unified consumption of products, which enables an increased market share of cultural products. However, at the moment, cultural industry and commercial industry are becoming increasingly mixed, to the point that radio and television, film, music and book publishing are also classified as commercial industries. In this way, the combination of culture and commercial industries has become a fairly "democratized" vehicle of popular culture. Which, through the packaging of cultural and art festivals and carnivals, has now been given the title of a cultural renaissance. In this way, "culture" and "industry" will be brought closer together. A trend which can "culturalize" industry and, on the other hand, "industrialize" culture and art (Chen, 2002a, b). The development of foreign creative industries has served as a reference point that Taiwan's cultural industry can learn from in the future.

2. Development trend of domestic cultural and creative industries

In the 1980s, the Council for Cultural Affairs advocated "community development". The cultural heritage generated in the history of regional cultural development, such as historic sites, folk art streets, folk art, festivals and festivals, are all based on the spirit of overall community building. Through extensive communication between local residents and experts, adapting measures to local conditions and adhering to the spirit of local culture, community residents can take initiative to participate in the planning and implementation of turning the locality into a brand-new local tourism resource. At the same time, it can also be regarded as the development of local culture and the build up of the locality's self-image. Not only that, the overall development of the community also impacts the local traditional handicraft industry. Since Taiwan has been mainly export-oriented for more than 40 years, its local craft culture lacked the opportunity to build up an image of its product. Therefore, in order to reverse the predicament of a dwindling market which the craft industry is facing, the Council for Cultural Affairs, under the general trend of community development, has adopted the concept of "cultural industry", to revive the memory of local residents and further infuse local culture into their traditional crafts. Projects which attempt to "industrialize culture, culturalize industry", are typically initiated by local entities, and begin in smaller areas before developing and advancing into larger areas from point to point to create the cultural characteristics of national industries (Weng, 1995). Therefore, the condition of the development of the local cultural industry makes the status of "product" and "information" interchangeable. The acquisition of the historical and cultural information regarding the industry becomes the main demand and focus, while cultural and creative products take a back seat. The desire to purchase cultural and creative products is only generated when consumers are interested and relate to the cultural importance behind the product. Therefore, the local cultural industry attaches great importance to the emotions generated through interpersonal interaction, a business practice which is quite different from most other industries (Hwang, 2001). As the Danish futurist Rolf Jensen said: "When we shop, we are actually looking for stories, friendship, care, lifestyle and character within the product, we are actually buying emotions (Feng, 2002)." The cultural and creative industries attempt to extradite the industrial bodies that originally belonged to the cultural industry and local cultural industries into the category of creative industries with a concept of cultural diffusion and industrial obscurity. Using "creative design" as the magic of turning stone into gold, these industries will seek the greatest cultural and economic benefits through the promotion of culture and art along with various commercial mechanisms. While also focusing on the importance of popularity and essence in the promoting of culture and art, to fully unleash the greatest common divisor in cultural economics (Chen, 2002a, b).

Taiwan is currently experiencing a change and crisis in the concept of cultural policy. From the article Economic Discourse in Cultural Policy, written by scholar Wang Li-Jung, the economic discourse in cultural policy has been continuously strengthened, and there is a possibility of close integration with economic and industrial policy. The emphasis of cultural policy transitioned from aesthetic and societal value to pure economic value. As the scope of cultural concepts continues to expand, the boundaries which have come to define these concepts continue to blur; whether it be art, entertainment, traditional elite, popular culture, extravagance, popular art, or public sponsorship, the distinction between the commercial and culture industries has gradually disappeared (Wang, 2003, 2005). The objects of cultural policy concern have expanded from traditional high-level culture and cultural industries to creative industries and so on. Judging from these trends, cultural policy is an unstable concept, whether it lies in the legitimacy of the discussion, the emphasis on value, or the definition of culture, each factor changes with social and historical changes, and presents different policy implications. Today, economic value and market discourse have become the slogans of "rationality" in cultural policy to win public subsidies and government support, but there are some crises that we must be alert to: whether the overemphasis on economic value is crowding out other values of culture and art, such as whether or not it is worth it, for the purpose of market opening or industrial development to jeopardize cultural diversity or the protection of cultural rights? While the concept of culture is constantly expanding and all-encompassing, what happens when certain important aspects are eroded and forgotten? What we need to think about is not whether the future cultural policy is still called "cultural policy" or "economic policy", but whether other values of culture and art are confused and ignored, as well as whether or not the role of culture in everyone's life and life has changed (Wang, 2003, 2005).

Compared with the experience of other countries, Taiwan obviously has a different development context. For example, in its attitude towards the "cultural industry", the cultural industry has always been excluded from cultural policies in the past. And it is worth considering whether it has caused obstacles and gaps in the industry's current development. Due to the policies of community building, overall community development and local creation (Wang, 2022), the accumulated special experience in "cultural industry" should be helpful for the promotion of the "cultural and creative industry". Therefore, when Taiwan proposed the concept of "cultural and creative industries" from the West, on one hand, it should clarify the social and policy context from which this concept came from, and on the other hand, it should compare the different experiences of the two to find a development orientation and direction suitable for Taiwan (Wang, 2003, 2005).

3. Policy comparison between cultural industry and cultural and creative industry

There is an obvious gap between "cultural industry" and "cultural creative industry" in Taiwan's cultural policy. This gap is particularly apparent in Taiwan's definition of culture and the values which it emphasizes. "Cultural industry" tends to think about the position of culture in terms of its implications in traditional and local life, as well as the process of human evolution and advancement; while "cultural creative industry" blurs the boundaries of culture, and views all products with cultural symbolic meanings to qualify as part of Taiwan's culture. In addition, "cultural industry" has the backing of a high degree of local support and community recognition, and it is not entirely viewed in terms of its pure economic value; in contrast, the "cultural and creative industry" is more like an industrial policy (Wang, 2003, 2005). Created by Chen Qinan, then chairman of the Council for Cultural Affairs, the following two tables highlight the differences between the two industries: Table 3.3 "Cultural Industry" with Table 3.4 "Cultural and Creative Industry".

From the implementation experience of the cultural industry in 1995, the knowledge economy in 2000, the cultural and creative industry in 2002, and the first year of local creation in 2020, it seems that the cognition and connotation of Taiwan's cultural policy have changed drastically in the past 30 years (Fig. 3.33). The field of expertise has shifted from the appreciation of high-level traditional art or the significance of educating citizens to pertinent issues such as cultural regional development and the development of creative industries. The vocabulary used has also changed

Table 3.3 Policies on cultural industries in Taiwan (Wang, 2003, 2005)

Definition	Comparison of the two industries				
	Definition	Area of reference	Purpose of the policy	Emphasized values	Definition of culture
Cultural industry policies (close to Community Development)	An industry which has local characteristics as well as local economic benefits, marked by its focus on culture as the core for development	Local tourism, crafts, maintenance of cultural assets, preservation of settlements, media, art of living, agriculture and fisheries, etc.	The development of cultural and artistic activities themselves and their products as local industries, or the use of existing and newly developed cultural and artistic activities and products as incentives, leads to projects of community building and restructuring which revitalize each locality	Community, identity, tradition and individuality. (cultural value and social value)	Focus on local traditional culture, excludes cultural industries. The cultural industry relies on creativity, individuality, local tradition and specialties, and even the originality and creativity of craftsmen or artists, emphasizing the usability and spiritual value of products

Table 3.4 Policies on creative industries in Taiwan (Wang, 2003, 2005)

Definition	Comparison of the two industries				
	Definition	Area of reference	Purpose of the policy	Emphasized values	Definition of culture
Policies regarding cultural and creative industries (closer to a knowledge economy)	The creative industry consists of three core concepts: the creation and publication of fine art; the applied art built on the core foundation of culture and art	Ten areas: visual arts, music and performing arts, crafts, design industry, publishing, television and radio, film, advertising, cultural performance facilities, leisure software	Increase the employment population within the cultural and creative industries; increase the market value of cultural and creative industries; improve the quality of culture in Taiwanese daily life; build a cultural industry with Taiwanese characteristics, improve creative styles and concepts; become a hub or platform for the creative industry in Asia	Economic viability, creativity	Blur the boundaries of culture; include all products which hold a significant symbolic meaning, including traditional high culture, culture industry, knowledge economy, and similar surrounding industries

from "cultural subsidy" to "cultural investment"; the purpose of cultural policy has also begun to emphasize "the increase in the output value of the cultural industry and the increase in the employment population (Wang, 2003, 2005)" (Fig. 3.2).

4. Industrial culturalization, cultural industrialization

Regardless of the ambiguity and problems encountered by the Taiwanese government in the formulation of cultural policies, local governments should continue to promote the cultural industry so that culture and industry are intensively combined. However, in the early stages of this promotion, the two areas "culture" and "industry" were still two terms that had different definitions, which are compared and explained in the table below (Table 3.5).

It is easy for culture to have commercial value when it becomes involved with industrialization, but the meaning behind the product of this involvement cannot move people. If the culturalization of the industry is done to obtain some additional value, such motives become clear to the public eye. And if the motives to squeeze

Fig. 3.2 The evolution of Taiwan's cultural policy

Table 3.5 Culture and industry: definition and merits/demerits (Lin, 1998)

	Culture	Industry
Definition	The attributes of culture are its concept of anti-consumption, and it's not quite functional nature, it is not quantifiable. Culture represents an essence which cannot be consumed. Culture is a presentation of an overall spiritual lifestyle and life itself, and anti-consumption is its essence	Industry exists because of consumption. Without consumption, there is no industry. Only when there is expected consumption behavior will a corresponding type of industry be developed
Pros	Concrete presentation revealed from the inside out	It is commercially viable and consumable
Cons	Emotionally touching, but not consumable	Has a commercially viable nature but no spiritual essence
Culture and industrial integration	Without culture, industry would have nothing to commercialize and produce	Without industry, culture will not be able to reach extraordinary heights and be touching to the public

additional value are clear to the public eye, then it is inevitable that the product will fall short of expectations (Lin, 1998). From 1995 to 2006 in Taiwan, from traditional local industries to cultural and creative industries, Taiwan's local governments initiated a cultural policy attempt to "industrialize culture and culturalize industry", developing from small areas to large areas, from point to plane, to establish key cultural characteristics within various national industries (Weng, 1995). "Industrial culturalization and cultural industrialization are ideas that Taiwan has put forward since the beginning of the promotion of community development. Industrial culture refers to the culture which defines an industry, that is, an industry (including agriculture, fishery and forestry) has been producing products (including agriculture, fishery and forestry) for many years. In the process of production, based on the experience of material selection and processing, creativity and quality control of product design and production, it has developed its own unique culture. There are many connotations worth learning, such as wine making is usually perceived as purely a production business, but the experience and knowledge accumulated over a long period of time make winemaking a precious culture (Hwang, 1998, 2001)." Therefore, in the process of local industry culturalization and cultural industrialization in Caotun, we can start with the local "traditional industry" of straw sandals, which was a local production business in the past. The accumulated experience, technology and knowledge aid in the discovery of the cultural identity of the industry, making straw sandal craftsmanship another example of an industry with a precious culture. At this time, the development of industrial culture, and the culture of straw sandal craft production became an object for visitors to learn and experience, thereby activating the cultural content of the industry, and finally "turning into a cultural industry", making health straw slippers a cultural industry commodity. See Fig. 3.3 for details.

Many places in Taiwan are facing transitional issues while in the process of implementing the initiative of "industrial culturalization and cultural industrialization". In the current climate, Taiwan's industries are either lacking cultural connotation, or that the industry has excessive commercial interests in bringing a product to the market for mass consumption, resulting in a weak connection with the local residents and culture. Furthermore, the products sold within the related local industries are also ignored. Finding a balance where the two sectors of industry and culture properly

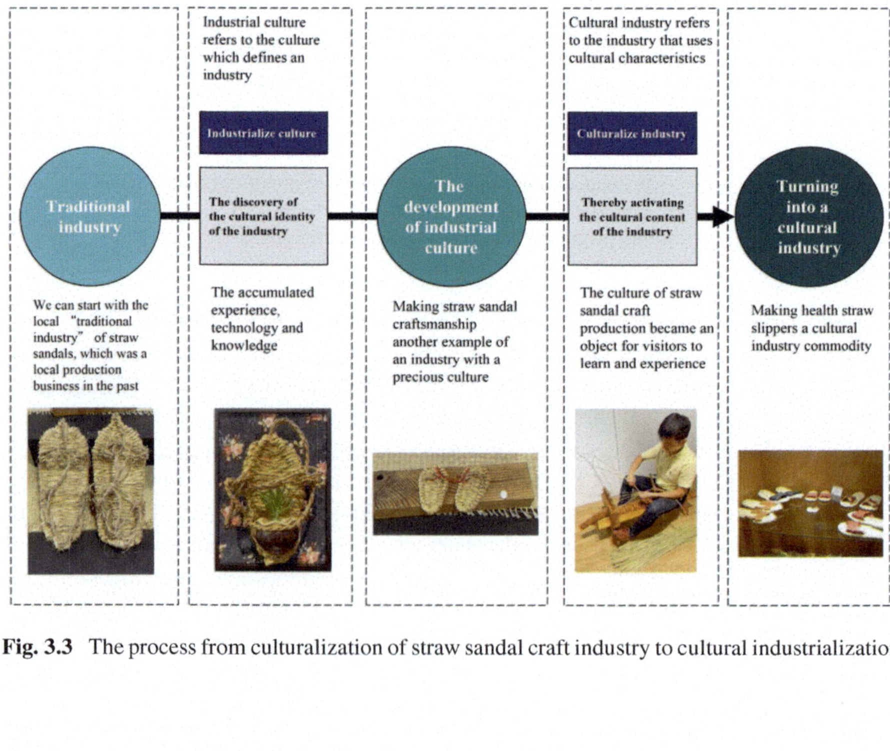

Fig. 3.3 The process from culturalization of straw sandal craft industry to cultural industrialization

complement each other is the greatest effort of "industrial culturalization and cultural industrialization".

5. Cultural tourism drives local industrial economy

Local industries must be culturalized, and culture must be industrialized, with the purpose of driving local tourism crowds. The relationship between tourism and cultural industries is rather unusual. Broadly speaking, both domestic and international tourism have cultural projects. Consumers may travel to a new place only for leisure desires or out of curiosity, but their experience in that place will undoubtedly take place in a cultural environment. Different environmental experiences are filled with a lot of cultural information, allowing the consumer to experience a sort of cultural baptism and alter their perception and recognition of culture. At the local level, these cultural industries may have a strong correlation with the tourism industry. On the one hand, they provide the tourism industry with a wide variety of products, and on the other hand, the tourism industry enables them to survive. More specifically, the tourism industry and the cultural industry are combined into one and occupy a niche market, which is the so-called "cultural tourism industry". National tourism is a combination of high-traffic, low-cost, packaged goods, while cultural tourism is a combination of low-traffic, high cost, and individualized (whether consumed in a group or individually), such as attending art festivals or visiting specific cultural sites, or extended cultural themed tours such as opera, literature, Aboriginal tribes and villages, archaeological remains, and art galleries. Although some cultural tourism enthusiasts claim that cultural tourism is a new phenomenon, it in fact has a long history. For example, the nineteenth century book The European Grand Tour is an existing and historical form of cultural tourism; and today, the rich and privileged still go on cultural tours for educational and self-improvement purposes. From the above sections, we can clearly see that cultural tourism will be closely linked with local cultural industries, and will also frequently interact with the culturally-rich lifestyles of local residents (Throsby, 2001). The cultural industry drives tourism, enables the country's economy to achieve economic growth through culture, and has a positive benefit on mass consumption in Taiwan.

6. Consumption status of activities related to cultural industries in Taiwan

With the improvement of the national economy, "cultural and creative industries" was listed within the 2008 National Development Plan, which proves that the country attaches great importance to the cultural economy. In fact, the implementation of cultural policies has enabled the gradual increase of Taiwan's consumption frenzy. According to data from the "Arts and Cultural Activities Information System" established by Taiwan's Ministry of Culture, Taiwan's cultural industry crowds and consumption have increased year on year over the past ten years. To illustrate this, the statistics of consumption and crowds from 2011 to 2020 are displayed, and the demographics of the population attending the various art and cultural activities in Taiwan are displayed to provide even more detailed data by year. Since 2011, the number of people participating in arts and cultural activities in Taiwan appears to have increased in the long run, especially in 2014, when the numbers reached their peak. Affected by the pandemic, in 2020 a total of 50,489 arts and cultural events were held across the country, with 186,532 thousand attendees. If measured by population size, in 2020, the average number of arts and cultural activities per 10,000 people is 21, and the average number of arts and cultural activities attended per capita is 7.92 (National Arts & Cultural Activities Information System of the Ministry of Culture, 2022).

It can be seen from the above data that there has been a prosperous development of cultural and artistic activities in Taiwan. Once the average income of individuals reaches a certain level, the demand for cultural and creative products will naturally increase, and this demand will ascend from basic physiological needs to psychological or even self-actualization needs. On average, 59 arts and cultural events are held in Taiwan every day. The meaning of this is twofold, first, it highlights the consumer appeal of Taiwanese cultural and creative products, and second, it explains the considerable profitability of the market for cultural and creative products. The mass production and circulation of consumer cultural and creative commodities gives the cultural and creative products their symbolic meaning. Developing people's consumption of cultural and creative commodities through consumption is also an aesthetic experience in people's leisure life. Through the hosting of cultural and artistic activities in the cultural industry, consumers are able to gain a new level of understanding towards the spiritual and cultural aspects of cultural and creative products. Tables 3.6 below show the statistics from the Ministry of Culture of Taiwan's National Arts and Cultural Activities Information System over the past ten years, as well as the demographics of the population attending arts and cultural activities across the country. This table highlights the prevalence of local arts and cultural activities in Taiwan, and the promotion of local cultural spending as a result.

7. Local industries and cultural and creative products

Once a local industry has cultural activities which drive the meaning behind or consumption behavior of products, cultural and creative products are formed. However, the forms of these products can be extremely diverse, and they can be both obvious and obscure. Therefore, cultural activities and their consumption are different from the consumption of general merchandise on the shelves in a store. Cultural symbols can be used as a screen to mask business operations. Exhibitions that appear to put public interests first, disguise the investment and profits made through business operations from these exhibits through the injection of cultural symbols. In his article Thinking about the craft industry: what else is needed besides the output value (Chen, 2002a, b), Yu states that cultural and creative industries focus on local crafts, and crafts can be divided into three areas: the first is "arts and crafts", the second is "craft products" and the third category is "community crafts". Crafts do not necessarily emphasize the artistic achievements of works, but carry the cultural characteristics of regional and community life. Therefore, it has evolved into the community craft industry concerned by the "community development" policy (National Culture and Arts Foundation 2004). It has been mentioned in previous sections that Caotun Town has been using the "craft town" model for many

Table 3.6 Statistics on the attendance of national art and cultural events (National Arts & Cultural Activities Information System of the Ministry of Culture, 2022)

Year	Number of individual activities	Attendance (Thousands)	Average number of arts and cultural activities (individual activity/10,000 people)	Average number of artistic and cultural activities attended per capita (attendance/person)
2011	59,300	200,634	26	8.64
2012	64,534	228,303	28	9.79
2013	66,389	254,333	28	10.88
2014	64,685	263,297	28	11.24
2015	53,908	247,187	23	10.52
2016	56,107	257,249	24	10.93
2017	58,195	263,037	25	11.16
2018	55,839	238,490	24	10.11
2019	57,246	255,389	24	10.82
2020	50,489	186,532	21	7.92

years as an important policy for industrial tourism, providing great influence to local industries and cultural creative products. Local industries cannot survive without cultural and creative products due to the fact that most areas of Taiwan operate under the traditional industrial economy. Under the impact of the modern knowledge economy, localities attempt to capture consumers through various cultural and creative industry related activities. At this time, local cultural and creative products have become an important tool to promote sales. If local industrial specialty products cannot quickly gain a foothold in the market, consistent innovation on the region's cultural and creative products must take place, in order for local specialty products to maintain economically beneficial to the locality.

3.3 Definition and Categorization of Cultural and Creative Products

Cultural and creative products are, as the name suggests, products derived from cultural characteristics, and mainly exist in conformance to prevailing economic interests. The commercialization of culture refers to the reduction of culture into a commodity under the capitalist commercial system, and the creation of a "can of culture", the product of a standardized production process meant for mass production created through minority culture and business elites. However, the commercialization of culture is not a reflection of the real needs of human beings, and the resulting products lack the principle of spontaneous creation. What it reflects is a mass culture dominated by the market economy, a culture which only reflects popular and mainstream social values. In economics, it is believed that commodity itself has intrinsic and exchange value and thus generates a price within a market (Lull, 1995). The biggest characteristic of this commodity is that it has characteristics that can be replicated in large quantities, and consumers will naturally act in their self-interest depending on their demand for the said commodity (Chen, 2002a, b).

The difference between cultural and creative products and general merchandise lies in the cultural implication hidden in and displayed from the goods, which is illustrated from the following definition: "since ancient times, human have known how to use stones, fire, bones, and tools to make living appliance, which become a product, and the product becomes a kind of goods after it gained economic value through commercial trade activities. The cultural elements inherent in the cultural and creative products are re-examined and reflected on, and they are displayed in the form of modern features through the design, and its function of satisfying the spiritual needs is explored. The difference between cultural and creative products and general merchandise culture lies in its Cultural Identity function (Ho et al., 1996)".

Therefore, in the activities of the International Straw Culture Festival in Caotun Town, if a product itself has meaning or behavior, it can be called a cultural and creative product, and it can take many forms, both obvious and obscure (Fig. 3.4).

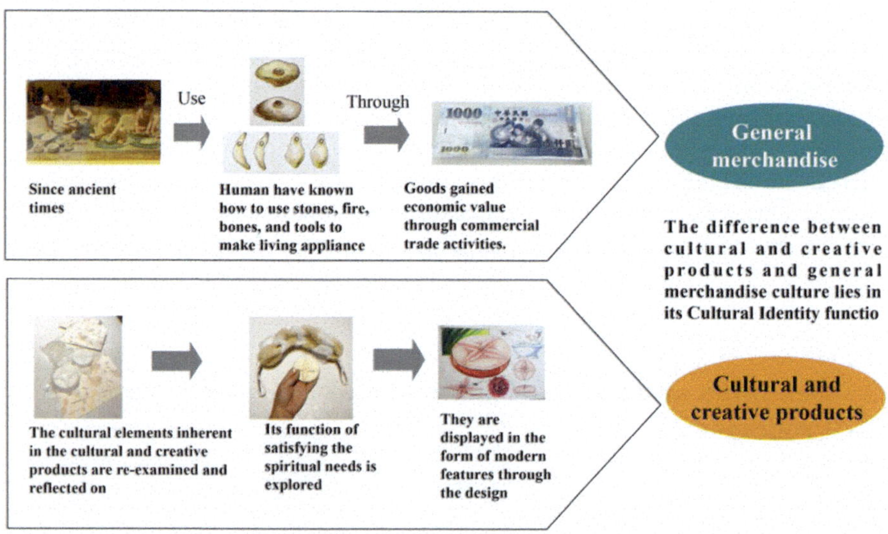

Fig. 3.4 The difference between cultural and creative products and general products (Ho et al., 1996; Wang, 2020)

3.3.1 Categories of Cultural and Creative Products

There are many cultural and creative commodities on the market, and the categories of cultural and creative commodities are divided into two types in the way of general consumers. One is the cultural and creative commodities in popular culture, and the other is the cultural and creative commodities with monopoly and unique properties.

1. Cultural and creative commodities under popular culture: Mainly in reference to packaging a cultural ideology into products that can be consumed by the general public. For example, Starbucks coffee, there are many franchises on the island, making it convenient and simple for consumers to buy Starbucks products (Chen, 2002a, b).
2. Cultural and creative products with monopoly and unique properties; such cultural and creative products cannot be quickly circulated and consumed in the mass market, so consumers may have to approach consuming these products with a "pilgrimage" mentality. For example, The Caotun cultural industry holds an event. The uniqueness of this event means consumers must spend time traveling and participating in person in order to consume this cultural and creative product. To briefly explain the difference between these two cultural and creative products, one is a cultural and creative product which can be taken with the consumer; the second is a cultural and creative product which cannot be taken with the consumer, time and money must be expended in order to experience or consume these products in person (Chen, 2002a, b).

The above section outlines the general differences between the two general types of cultural and creative products. Products are inherently identifiable, but depending on changes in the surrounding environment, products will also take on a different meaning (definition). Therefore, from the existing product categories, products can be divided into four additional categories, one is "cultural and creative products shaped by product consumption", the second is "personal consumption of private cultural and creative products", and the third is "collective consumption of publicly-owned cultural and creative products", and the fourth is "cultural and creative products with the characteristics of a consumption monopoly (Chen, 2002a, b)."

1. Cultural and creative products shaped by product consumption: the definition is to inject products into a system of symbols, whether these systems of symbols be cultural, spatial, language, etc. For example, when you go to a coffee shop to have a cup of coffee, you can enjoy and consume various symbols of that culture. In addition to selling tickets, exhibitions also provide some peripheral cultural products, such as clothes, mugs, postcards, etc., all of which extract and inject a cultural ideology into different products, making it a medium through which a culture can be brought to the public (Chen, 2002a, b). For example, Starbucks coffee (Fig. 3.5), IKEA (Fig. 3.6).
2. Personal consumption of private cultural and creative products: It is defined as privately owned cultural and creative products, but some products in this category are produced via a huge quantity of replication technologies, such as cultural and

Fig. 3.5 Starbucks coffee

Fig. 3.6 IKEA

artistic reproductions and the like. However, to only measure the value of these products from the perspective of their pricing completely ignores the characteristics of the cultural market, from the perspective of demand, consumers who maximize their utility and are unaffected by time constraints are favored in the market, along with the effects on a product's price in the market over time. For example, if an artist produces a piece of work, its true value cannot be immediately reflected in the market through its initial asking price. A work may have to go through the "drilling" of time and pass the "judgment" of critics before its true value is reflected in the market. A more practical example would be the value of a painter's work skyrocketing after his death, simply because he can no longer produce more works, making his existing works far more scarce in the market, as no new ones will be supplied (Chen, 2002a, b). Sales of the works of Chang Dai-Chien, a painter and calligrapher, increased substantially after his death due to an escalation of the scarcity of his works (Fig. 3.7).

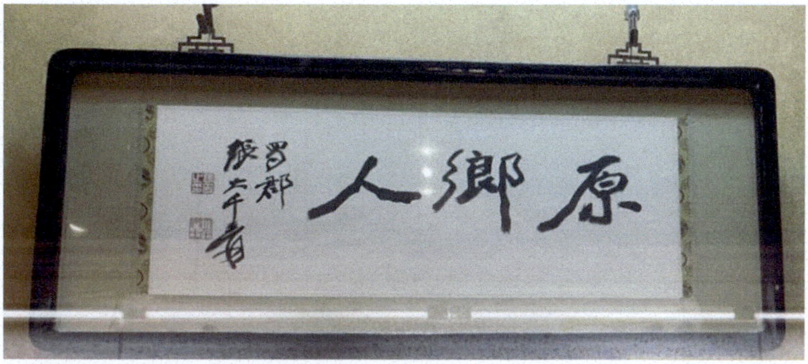

Fig. 3.7 Calligraphy by Chang, Dai-Chien (Chung, Li-Her Museum 2022)

3. Collective consumption of publicly-owned cultural and creative products: In terms of public-owned cultural and creative products, they cannot be privately owned, and they are impossible to evaluate their flow in the market by means of product supply. For example, if public goods in the cultural field are regarded as products, how should ticket prices be set to simply evaluate these products? What is the maximum amount one should donate to these public goods? Whether or not consumers are willing to make regular donation payments, etc. (Chen, 2002a, b). For example, antique charity sales, etc. (Fig. 3.8).
4. Cultural and creative commodities with the characteristics of a consumption monopoly: Defined by the operation of cultural and creative commodities where only the institutions or individuals who control the property rights can decide whether it can be directly or indirectly traded in the market, such as historical relics, particular lands which grow a specific crop, original works of art, etc. Such a monopolistic business model actually conforms to the core meaning of the cultural activity itself. For example, consumers will go to the local cultural industry for leisure and sightseeing. Through the act of tourism, consumers pay the cost of sightseeing or consuming local products, all of which provide an economic benefit to the local culture industry and qualify as the consumption of cultural and creative products. In fact, among these fees charged, it is because of the particularity and uniqueness of the local cultural industry that compels consumers to spend this fee of "pilgrimage" from their wallets (Chen, 2002a, b). For example, cultural and creative products of the National Palace Museum in Taiwan (Fig. 3.9), cultural and creative products of the Yilan BaiMi clogs community (Fig. 3.10), cultural and creative products of various ethnic groups in Taiwan (Fig. 3.11), and cultural and creative products of local health slippers in Caotun, Nantou (Fig. 3.12).

Fig. 3.8 Blue-and-white porcelain (The Shanghai Museum, 2017)

Fig. 3.9 Cultural and creative products of the National Palace Museum in Taiwan (The National Palace Museum, 2022)

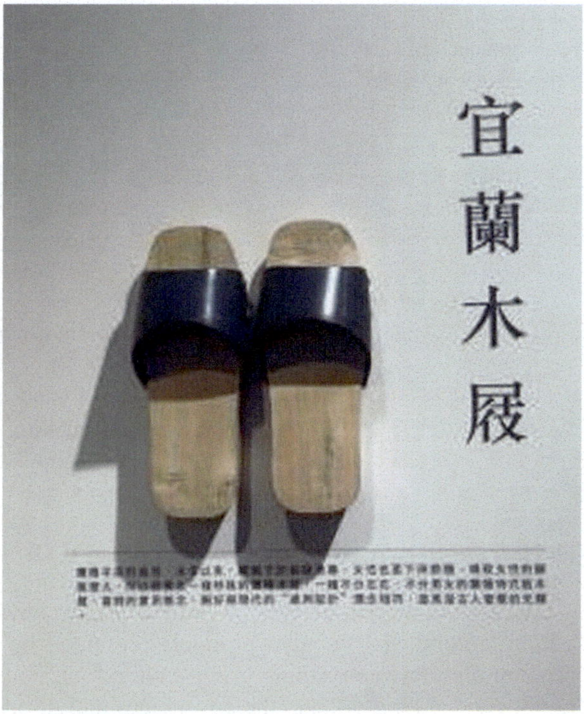

Fig. 3.10 Cultural and creative products of the Yilan BaiMi clogs community (Wooden Clog Museum, 2022)

Fig. 3.11 Cultural and creative products of various ethnic groups in Taiwan (Wang, 2018)

3.3.2 Analysis of Cultural and Creative Commodities with Monopolistic Consumption Characteristics in Taiwanese Museums, Communities, and Local Areas

From academic literature, we can clearly learn that cultural and creative products with monopolistic consumption characteristics have similar and different qualities as museums, ethnic groups, communities, and local cultural and creative products under the impact of Taiwan's cultural and creative industries. These unique qualities

Fig. 3.12 Cultural and creative products of local health slippers in Caotun, Nantou (Caotun Straw Crafts Museum, 2020)

cannot be imitated or replicated elsewhere, and create the monopolistic nature of these products.

Based on domestic successful cultural and creative product design, such as the fashion cultural and creative products presented by the Taiwan National Palace Museum and Taiwan's ethnic group cultural and creative products, Yilan County Baimi wooden clog village, the advantages and disadvantages of the local cultural and creative products of Caotun Township, Nantou County were explored to find out the problems in design and promotion. The following listed and described the features of Taiwan's cultural and creative product:

1. The fashion cultural and creative products of the Taiwan National Palace Museum

The National Palace Museum (Figs. 3.13 and 3.14) occupies a leading position among domestic museums and ranks as one of the important world-class museums. Facing the impact of cultural creative industry tide, the Museum has become not only just a place to display the antiquities, it needs to monopolize the cultural and creative products in the market. For cultural innovation and transformation, we invited various designing teams, including that of the well-known Asustek, college students majoring in design, and famous design studio to design and develop cultural and creative products. We developed a series of cultural and creative products related to the Palace Museum. The designers made full use of the cultural resources in the Museum, such as history and stories, to convert our daily necessities into products. The full range of cultural and creative products entailed the elements of symbolism, the conversion of commodity function, and cognition of historical stories. The followings are goods sold by the Palace Museum (Taiwan Design Research Institute, 2005).

In this study, five case studies on the design of cultural and creative products of the National Palace Museum in Taiwan are listed, Xi Bodkin with dragon pattern, late Warring States Period to early Western Han dynasty, circa 275–141 B.C.E. (Fig. 3.15). Using the symbol of the Xi Bodkin with a dragon pattern the designer

Fig. 3.13 Exhibition hall—main building (The National Palace Museum, 2022)

transforms the dragon's tail into a letter opener. The dragon's whiskers under its head have the function of opening a bottle, the combined functions of this product give it the name "Dragon opened the letter opener" (Fig. 3.16). This product is an aluminum product with a strong magnet inside, which can be attached to a metal surface and used as a hanging product (The National Palace Museum Shop, 2022).

Meat-shaped Stone, Qing dynasty, AD1644-1911 (Fig. 3.17). The craftsman who made this meat-shaped stone took the rich natural resources of this stone and carved it with great precision, and then the skin was stained. This process resulted in the appearance of skin and lean and fatty layers of meat, the veining and hair follicles making the piece appear even more realistic (The National Palace Museum, 2022). The designer transformed the shape, color, quality and texture of the meat-shaped stone from the Qing dynasty into a product known as the national treasure book clip: meat-shaped stone. A product which can hold documents, small paper cards and mini book clips, works as a great office utility (Fig. 3.18).

Pillow in the shape of a recumbent child with white glaze, Ding ware, Northern Song dynasty AD960-1127 (Fig. 3.19). Height: 18.8 cm, diameter at base: 31 × 13.2 cm. The design is both life-like and alluring, with the child, clearly in excellent health, wearing a suit of clothing including a long brocaded top, lying on a mattress, legs crossed behind (The National Palace Museum, 2022). Paperweight-Pillow in the Shape of a Recumbent Child, made of unsaturated polyester resin with a white porcelain appearance (Fig. 3.20).

Fig. 3.14 The National Palace Museum (2022)

Fig. 3.15 Xi Bodkin with dragon pattern, late warring states period to early Western Han dynasty, circa 275–141 B.C.E. (The National Palace Museum, 2022)

Fig. 3.16 Dragon opened the letter opener (The National Palace Museum Shop, 2022)

Fig. 3.17 Meat-shaped stone, Qing dynasty, AD1644-1911 (The National Palace Museum, 2022)

Fig. 3.18 National treasure book clip: meat-shaped stone (The National Palace Museum Shop, 2022)

Fig. 3.19 Pillow in the shape of a recumbent child with white glaze, Ding ware, Northern Song dynasty AD960-1127 (The National Palace Museum, 2022)

Jadeite Cabbage, Qing dynasty, AD1644-1911, Length: 18.7 cm, width: 9.1 cm, thickness: 5.07 cm, Some have surmised that this piece was a dowry gift for Consort Chin to symbolize her purity and offer blessings for bearing many children. Let's also not forget the two insects that have alighted on the vegetable leaves (Fig. 3.21). They are a locust and katydid, which are traditional metaphors for having numerous children. This item converts the Jadeite Cabbage into a product called the Jadeite Cabbage Digital Camera using modern materials such as resin, acrylic, wood and brass (Fig. 3.22). The half-chartreuse, half-white camera surface was created using fluid resin technique to resemble the pattern of jade. It recreates the color of the famous sculpture, Jadeite Cabbage. The internal structure of the camera is faintly visible through the semi-translucent resin surface (The National Palace Museum Shop, 2022).

Jade Pi-Hsieh, auspicious beast, Eastern Han dynasty AD25-220 (Fig. 3.23). The "pi-hsieh" is a mythological creature commonly thought to be able to ward off evil forces with its magical powers. In fact, its name means "to ward off evil" in Chinese (The National Palace Museum, 2022). Paperclip Holder Paper Weight-Pi-hsieh made of unsaturated resin with a magnet embedded in the midsection of Bixie (Fig. 3.24). The item is designed to hold paper clips with its magnetic design, and could also be used as a paper weight (The National Palace Museum Shop, 2022).

Fig. 3.20 Paperweight-pillow in the shape of a recumbent child (The National Palace Museum Shop, 2022)

Therefore, the injection of symbolism, functionality, history, narrative and uniqueness, etc. into the cultural and creative products of the National Palace Museum in Taiwan is quite obvious. And the designs make use of design elements such as shape, color, material, texture and function to translate historical relics into daily necessities, bringing the public closer to the National Palace Museum.

2. Clog-based cultural and creative products of BaiMi Wooden Clog Village, Yilan County

The Bai Mi wooden clog industry in Su'ao Township, Yilan County is facing the crisis of imminent shutdown, but two clog masters remaining in the community are still engaged in the mission of passing down local traditional craftsmanship. Baimi Wooden Clog Village, previously known as White Rice Urn, was named after its local production of rocks shaped like rice grains. These local national treasure-level masters and local residents often chat after meals about their respective memories regarding their experiences secretly cutting down the trees surrounding BaiMi village during the period of Japanese occupation to manufacture wooden clogs (Fig. 3.25). The phrase "white rice urns, make the clogs, steal the wood, and capture the forest" (Taiwanese) is permanently ingrained in the memory of many local residents (National Culture and Arts Foundation, 2004), and is also printed on the packaging of the village's products (Fig. 3.26).

Fig. 3.21 Jadeite Cabbage, Qing dynasty, AD1644-1911 (The National Palace Museum, 2022)

The BaiMi Wooden Clog Cultural Center began with community development, and developed its local communities and their spirit based on the local culture (). Furthermore, the center used the development of unique local industries and culture infused packaging as a means of creating a high-quality example during the first wave of overall community development in Taiwan. In fact, during the era of Japanese occupation and under the influence of Japanese culture, it became popular to wear wooden clogs in the region. BaiMi Urn (the local ancient name of BaiMi Village) was rich in the tree "Jiang Moushu (Fig. 3.27)", which was used as the base material in the production of wooden clogs. Thus, the clog industry gradually developed and became a key supplier for wooden clogs in Taiwan. With The assistance of the Council for Cultural Affairs and the Small and Medium Enterprises Administration of the Ministry of Economic Affairs, BaiMi Village has been revitalized and driven by its cultural industry, striving to become a community of life-long learning. With continuous personnel training and environmental renovation, it has become one of many "local cultural centers" developed by the Council for Cultural Affairs (National Culture and Arts Foundation, 2004).

There is an entire three-story Wooden Clog Museum in the village (Fig. 3.28). The third floor consists mainly of conference rooms, clog making exhibits, and restaurants (places to eat and drink), where visitors can gain a better understanding regarding

Fig. 3.22 Jadeite Cabbage digital camera (The National Palace Museum Shop, 2022)

Fig. 3.23 Jade Pi-Hsieh, auspicious beast, Eastern Han dynasty AD25-2201 (The National Palace Museum 2022)

the origins of wooden clog culture (Figs. 3.29 and 3.30). On the second floor, there are exhibits which introduce the history of BaiMi Village and display the works of wooden clog masters (Figs. 3.31, 3.32, 3.33 and 3.34). The first floor is mainly for the display and sale of clogs, where visitors can buy ready-made clogs or clog paraphernalia, or order a pair of custom clogs on the spot (Fig. 3.35). In addition,

Fig. 3.24 Paperclip holder paper weight-Pi-hsieh (The National Palace Museum Shop, 2022)

Fig. 3.25 White rice village wall (Wooden Clog Museum, 2022)

at a specially set up DIY experience section, audiences can learn to paint clogs and make them by themselves, or create sand paintings utilizing a local resource known as “white sand” (Figs. 3.36, 3.37, 3.38, 3.39 and 3.40).

The overall theme is clear, and the atmosphere of the store also emphasizes the nostalgic style of the olden days, using a large quantity of wood and tools in its

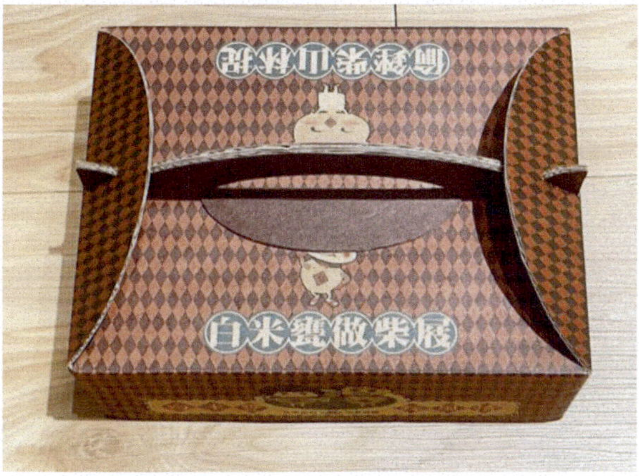

Fig. 3.26 Wooden clog box (Wooden Clog Museum, 2022)

Fig. 3.27 Schefflera tree (Wooden Clog Museum, 2022)

decorations to evoke the spirit of local daily life. In terms of product performance, it is even more remarkable. More than ten cultural and creative products have been derived from the single theme of "clogs"; and with new hand-made products such as leather carvings and sand paintings, the image of a museum dedicated to common people situated among folk lands can be seen with a bird's eye view, which can be described as the store culture of a successful local industrial museum (National Culture and Arts Foundation, 2004).

Fig. 3.28 Wooden clog museum (Wooden Clog Museum, 2022)

Fig. 3.29 The 3rd Floor briefing room (Wooden Clog Museum, 2022)

Fig. 3.30 The 3rd Floor balcony (Wooden Clog Museum, 2022)

It is clearly recognized in academic literature that Taiwan's community cultural industries have quite profound stories, such as Yilan's BaiMi Village and its clog-related cultural and creative products. Through the initiative of community cultural guides and community residents, who elaborate upon the rich historical stories of the community. In terms of cultural and creative products, stories from the community add value to the culture creative product itself and give additional meaning to the "spirit" behind the product. BaiMi clogs take this process one step further through the narrative guide and the personal experience of tourists' clog making, to create a personalized cultural and creative product for tourists, so that the performance of cultural consumption numbers of the local community continues to rise steadily. The connotation of Yilan BaiMi clogs-related cultural and creative products is mainly based on a narrative level. Local craftsmen paint strong symbols, and infuse local history, functionality, and other recyclable local industrial materials to highlight the basic function of wooden clogs and the uniqueness of the community spirit.

3. Taiwan ethnic groups' cultural and creative products

Taiwan, an island with colorful marine culture, comprises many ethnic groups and different cultures. In recent years, various ethnic groups hold sightseeing cultural festivals independently. In order to pass on their own ethnic cultures, they hold cultural activities in the form of competitions, and quantify the cultural and creative products as awards for the competitions to maintain the economic benefit on one

Fig. 3.31 The 2nd Floor clog making process exhibit (Wooden Clog Museum, 2022)

Fig. 3.32 The 2nd Floor large-scale clog exhibit (Wooden Clog Museum, 2022)

Fig. 3.33 The 2nd Floor tandem clog exhibit (Wooden Clog Museum, 2022)

hand and create self-brand products for their own ethnic group on the other hand. Sanshan Guowang, Lord of the three mountains, is an important god set of popular religion in Taiwan. Lord of the three mountains is a collection of three mountain gods who originated from Chao zhou, Guang-dong province. The three mountains are Mountain Jin, Mountain Ming, and Mountain Du. San-shan is their collective name. When the ancestors immigrated to Taiwan from mainland China, they took the "Incense Burner" from the Sanshan Guowang's temple as protection. When they arrived safely, the ancestors built a temple in Taiwan to worship and thank the Sanshan Guowang for his protection. Therefore, this product is to convert the symbol of "Incense Burner" from the Sanshan Guowang temple into a practical product. The cover and the body of the box shape the look of the Sanshan Guowang Temple. At the same time, snow-white porcelain clay and celadon glaze are used to represent the mountain god of the Sanshan Guowang from nature. This product means that it can bring protection and good luck from Sanshan Guowang, and is named "Lucky Box" (Wang et al., 2021) (Figs. 3.41, 3.42, 3.43, 3.44 and 3.45).

The fragrance of the magnolia coco is refreshing. It blooms at night, with a pure white appearance. The flower does not look fancy, appears relatively modest, and hasn't the slightest hint of excessive beauty. In the eyes of the Hakka people, the magnolia coco's white petals are wrapped in layers of green skin during the day,

Fig. 3.34 The 2nd Floor clog wall display (Wooden Clog Museum, 2022)

Fig. 3.35 Wooden clog (Wooden Clog Museum, 2022)

just like how a working woman wraps herself in layers of headscarves to avoid the scorching sun during a hard days' work. Releasing its light fragrance at night, the flower is akin to the working woman at the end of a long workday, unloading her burdens and removing her headscarves to exude an elegant feminine fragrance, revealing her gentle face as she spends the remaining hours of her day with her

Fig. 3.36 The 1st Floor decorative clog exhibit (Wooden Clog Museum, 2022)

family. In this sense, the magnolia coco is not only a symbol which represents Hakka women's industriousness and simplicity, but also serves as a representation of the tenacious life attitude of modern women (Li et al. 2008; Grandpa Wen Caitang, 2022). The shape of the wooden cutting board of this product presents geometric and symmetrical magnolia coco, the cutouts designed to suspend the cutting board placed at the handle display the added value of Hakka culture. The cutting board of this series of products is used for cutting, placement, storage, and decorative purposes.The design concept of this product has the intention of dispersing fragrance and sharing delicious taste, creating a comfortable life and improving the quality of one's soul (Figs. 3.46 and 3.47).

4. The interpretation of the straw and creative products of Caotun Township

Caotun Township, situated in the middle of Nantou County, Taiwan, was originally named "Cao Xie Dun" because during the period between Emperor Qianlong and Emperor Jiaqing, the farmers going up and down Lugang and Puli, porters, travelling merchants came to the Cao Xie Dun, abandoned their old straw sandals and put on new ones, and the old sandals formed a pier of sandals. The Cao Xie Dun today is called "Caotun" and the time is most likely to conform to the literature archaeological result. According to the Caotun Township Chronicle, there were three major ports in

Fig. 3.37 The 1st Floor Clog making tools exhibit (Wooden Clog Museum, 2022)

Taiwan in the olden days: Yifu (Tainan Prefecture), Erlu (Lugang), and Sanmenga (Wanhua). With Lugang claiming the title of the largest cargo port in central Taiwan at the time. The reason behind the development of these three ports, is that in the Qing Dynasty and the early days of Japanese occupation when the material was scarce, in order to support the livelihood of the family, early settlers divided time between farming and transporting goods, and thus went to Lugang to carry goods to the Puli Village. Porters relied on their shoulders to carry goods along the route, passing through Lukang, Changhua, Kuaiguan, Shekou, Xianzhuang, Beitou, and Cao Xie Dun, and enter Puli in the east, rest halfway, stay overnight, and replace their worn-out straw sandals (Compiled by Caotun Town Office, 1986).

(1) The origin of the name Caotun

Caotun Township, the largest town in Taiwan, was formerly known as “Cao Xie Dun”. This place name likely had appeared 23 years before Qianlong, and because of the appearance of this historical document, there are many different variations as to the true story of how Caotun got its name. As for the origin of Caotun’s place name, it can be roughly divided into the following five theories:

A. It is said that three thousand troops traveled to fight the aborigines at Douwei Longan, and they abandoned their shoes at this place, creating a mound. In

Fig. 3.38 The 1st Floor DIY tool cabinet/table (Wooden Clog Museum, 2022)

addition, it is also said that Zheng Chenggong's troops were going to enter Puli to attack the local aborigines tribes. Originally, they planned to pass through Guoxing, but because of cholera in the region, they changed their course and passed through Cao Xie Dun, where the troops rested and replaced their straw sandals. The straw sandals that were discarded piled up into little mounds, thus giving the place the name of Cao Xie Dun (straw sandal mound). Due to the fact that Zheng Chenggong died in Taiwan after only a year and four months on the island, during nine months of which he fought against the Dutch, it would have been impossible for him to lead his army northward, where the township is. Furthermore, the one who fought the aborigines at Douwei Longan was Zheng Jing according to *Pihai Travelog*. Thus, the "Zheng story" is naturally refuted (Compiled by Caotun Town Office, 1986).

B. The same as the above, this story also claims Guoxing is Zheng Jing because he was the one who fought the aborigines at Douwei Longan. Ino Kanori believes that it may be the Atayal ethnic group in the upper reaches of Beigang River; Lian Yatang believes that it could have been aborigines groups in regions north of the Dajia River; Mr. Weng Jiayin believes that the battle took place to the east of Dadu Terrace.

C. It is said that Lin Shuangwen was defeated in the region and led his troops past modern day Caotun, where they discarded their straw sandals, again creating the aforementioned small mounds of shoes. This story comes from Lin Shuangwen's

Fig. 3.39 The 1st Floor clog display case (Wooden Clog Museum, 2022)

failed attempt to fight against the Qing in the 53rd year of Qianlong. His path of retreat passes through the area, and traverses the Huoyan Mountain Road. However, contemporary records already contain the name Cao Xie Dun, which appears at least 31 years before Shuangwen's defeat. The name came before the event, creating an obvious discontinuity error within this story (Compiled by Caotun Town Office, 2005).

D. During the Qianjia period, business travelers and porters who settled the Tucheng Plain from the time of Daoguang to the beginning of the Japanese occupation, traveled to and from the Puli Basin, and all travelers used Caotun as the key point of entry and exit. Due to its location, it was an ideal place to rest, resupply, and replace straw shoes, the abundance of which gave the place its name (Compiled by Caotun Town Office, 2005).

E. This theory was put forward by scholar Chen Zhesan. He suspected that the Cao Xie Dun was named after the topography of the current Caotun Zhongshan Park, which appeared in historical documents from year 37 of the Qianlong period (Chen, 2002a, b). This statement is also written in the Caotun Township Chronicle. It is explained in terms of the general principle of place name formation. There are index place names first and then occupied place names. The terrain is named (the fifth theory), and it will gradually form the name of the place in the future. Therefore, if the theory's conjecture is correct, it can be explained that before Qianlong, the "Cao Xie Dun" was an indicator place name.

Fig. 3.40 The 1st Floor large scale painted clog exhibit (Wooden Clog Museum, 2022)

Fig. 3.41 The box can hold candy and nuts (Wang, 2007)

Fig. 3.42 The pens box (Wang, 2007)

Fig. 3.43 The box can hold other items (Wang, 2007)

Fig. 3.44 The tea box (Wang, 2007)

Fig. 3.45 Lucky Box for Shansan Kings (Wang, 2007)

After the late Qianlong period, the Han people gradually entered, and the Cao Xie Dun slowly became the occupied place name. The fourth theory regarding small mounds formed by discarded straw shoes was likely developed during this time of Han immigration. As for the latest point in time when Caotun was given its name, the records give no indication. The place name of Cao Xie Dun was

Fig. 3.46 Hakka wooden cutting board concept sketch (Wang, 2007)

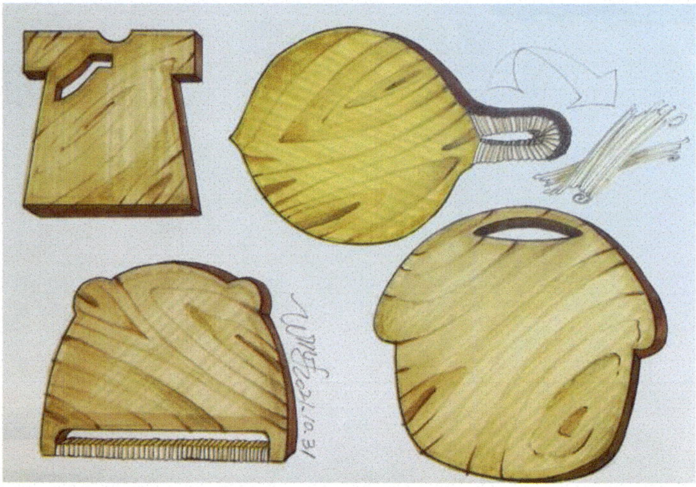

Fig. 3.47 Hakka Magnolia Coco wood cutting board (Wang, 2007)

used until October 1920 (the ninth year of Taisho, the ninth year of the Republic of China), when the Japanese adjusted the local administrative area and implemented the system of prefecture, city, and street to aid in local governance, thus abolishing the Beitou Fort, the Japanese administrative branch office in Cao Xie Dun, however, the branch office eventually changed because Beitou Street was destroyed by Japanese soldiers, and was moved back to Cao Xie Dun. The place name was simplified and renamed Caotun Village. In 1938, it was upgraded to Caotun Street. After the restoration of the province, it was changed to Caotun Township in 1950, the name which is used to this day (Compiled by Caotun Town Office, 2005).

(2) The evolution of "straw sandal" products in Caotun Township

The place name of Caotun Township is closely related to straw sandals. More than that, in the cultural industry, the image of straw sandals has transformed from the traditional impression to the current concept of indoor health straw slippers. Table 3.7 below discusses the evolution of Caotun's "straw sandal" products. From this table, we can better understand the grass sandal industry in Caotun. After the "industrial culturalization, cultural industrialization" of local industries, the evolution of straw sandal products went from traditional, funeral, crafts to indoor health straw slippers, which now stand out as a unique local cultural and creative product.

The table above describes the evolutionary history of "straw sandals". After early settlers from the mainland immigrated to Taiwan during the Qing Dynasty, the traditional straw sandals became necessary when traveling from Lugang through to Puli for the purposes of trading, felling, fishing, walking, and combat. After the Restoration of the Republic of China, the decline of traditional straw sandals in the 40 and 50 s forced the industry to transition and instead focus on meeting the need of straw sandals for funerals. With the development of the plastic industry, plastic-based shoes became common; at this time, straw sandals gave the older generation a negative impression as an unlucky industry. Until about the 80th year of the Republic of China, when the local government of Caotun, the Taiwan Craft Research and Development Institute, and the Cao Xie Dun local cultural and educational association vigorously promoted the spirit of a "craft town", and brought the straw sandals products out of the "dark ages" to become important local traditional crafts and decorative souvenirs. However, this was not the ultimate goal of the Caotun locality. With society prospering, people have more diversified demands for local industrial commodities as well as greater options to choose from; the local cultural group Cao Xie Dun Local Cultural and Educational Association has tried its best to change the image and function of straw sandals. Around the 1990s, the state promoted the cultural and creative industries, and launched the concept of indoor health straw slippers; a product that meets the needs of life and has local commemorative value.

(3) Straw sandals weaving tools and production process

People used to say: "If you want to do a good job well, you must also have access to the best tools." It is not easy to make a pair of straw sandals, and you must have a complete set of tools before you can weave memorable "straw sandals". In the straw workshop of Caotun, tools such as chair bars, straw sandal rakes, straw sandal rakes, straw sandal grooves, straw sandal belts, straw mallets, straw, kenaf and wax are placed to allow tourists to personally experience the craftsmanship of large and small straw sandals and attract further tourists, resulting in an increased understanding of the cultural identity of local cultural and creative products.

Figure 3.48 shows the main tools for weaving straw sandals, which are arranged in sequence as 1. leaning bar; 2. straw sandal rake; 3. straw sandal rake; 4. straw shoe groove; 5. straw sandal belt; 6. straw mallet; 7. scissors 8. wax; 9. straw; 10.

Table 3.7 Evolution of straw sandals in Caotun

Evolution aspect of straw sandals	Evolution item			
	Traditional straw sandal	Funeral straw sandals	Craft straw sandals	Indoor health straw slippers
Time period	During the Qianlong and Jiaqing years of the Qing Dynasty, and the period of Japanese occupation	During and after the fifth decade of the Republic of China	Begun during the eight-sixth year of the Republic of China	Begun during the ninety-third year of the Republic of China
Market demand	Straw sandals broken on the road from Lukang to Caotun needed to be replaced before heading East to Puli	Traditional sandals were also purchased as funeral wear	Straw sandals became local traditional crafts and souvenirs	Add innovative and practical concepts to traditional straw sandals, enabling them to become a new cultural and creative product
Reason for transition	There were many straw sandal craftsmen in Caotun who relied on the demand for replacement sandals to maintain their livelihood	With the arrival of modern shoes, the use of straw sandals in daily life dwindled, thus the industry transitioned to meet funeral needs	Local governments, the Taiwan Craft Research and Development Institute, Cao Xie Dun Local Cultural and Education Association vigorously promoted the "craft town" spirit	Promote cultural and creative industries to transform the traditional craft of straw sandals into "healthy indoor straw sandals"
Contemporary usage	Chopping wood, fishing, walking, combat	Funerals, Taoist spirit medium's shoe	Souvenirs, small sandal decorations	Product which caters to life necessities

(continued)

Table 3.7 (continued)

Evolution aspect of straw sandals	Evolution item			
	Traditional straw sandal	Funeral straw sandals	Craft straw sandals	Indoor health straw slippers
Sample image				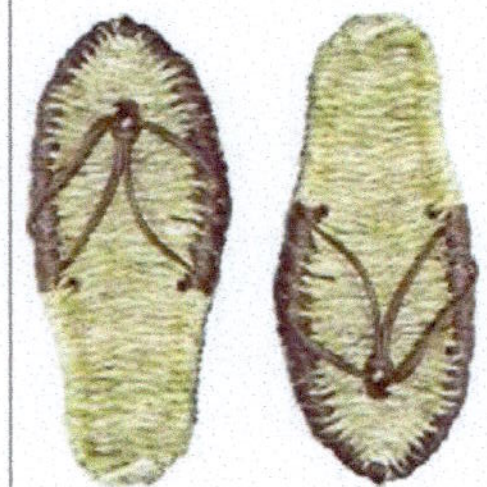

kenaf. The following Table 3.8 illustrates the production process of straw sandals. Following the order of, sorting the straws, selecting the straws and pounding the straws, rubbing the straws, tying the straws to the waist, straw sandals warp waxing, weaving (two up and two down), tightening to form the tip of the shoe, vamp weaving (one up and one down), measuring the position of front sides, making the front side (make four on each side), making the land god sides (three on each side), tightening the weaving, making the heel, tightening the heel, tying a knot on the heel, tightening the knot, trimming the vamp and the sole, wearing and tying the back heel and the tail rope (finishing the straw shoes); these are the complete steps for making sandals.

(4) Current status of local cultural and creative products in Caotun Township

Caotun Township is rich in cultural resources and well known for the traditional crafts of making straw sandals. Some cultural festivals are always held in the town, which brings business opportunities and aims to carry out the goal of "commercialization of local industrial cultural and creative products". But because the goods throughout the town lack unique features, it is relatively difficult for local residents and sightseeing consumers to identify with these goods. The local people actively retrieve the traditional crafts of sandals which have already downfallen, and commercialize the traditional straw sandals culture in the hope of bringing in local economic benefit and improving local culture and tourism industry in Taiwan. At the top of the best-sellers is "indoor healthy straw slippers", followed by the cultural and creative

Fig. 3.48 The main tools of straw sandal weaving

Table 3.8 Process of making straw sandals

(1)Sorting the straws	(2)Selecting and pounding the straws	(3)Rubbing the straws	(4)Tying the straws to the waist	(5)Straw sandals warp waxing

(6)Weaving (two up and two down)	(7)Tightening to form the tip of the shoe	(8)Vamp weaving (one up one down)	(9)Measuring the position of front sides	(10)Making the front sides(make 4 on each side)

(11)Making the land god sides (three on each side)	(12)Tightening the weaving	(13)Making the heel	(14)Tightening the heel	(15)Tying a knot on the heel

(16)Tightening the knot	(17)Trimming the vamp and the sole	(18)Wearing and tying the back heel and the tail rope (finishing the straw shoes)		

products which are woven by straw materials, among which the scarecrow pencils, various straw cards, straw knives account for the majority. See Table 3.9 (Caotun Straw Crafts Museum, 2005).

Compare the differences of the four cultural and creative products, we found that the cultural and creative products of the "National Palace Museum" and "Taiwan ethnic group" do better than that of the "Caotun Township" and "Yilan County Baimi Wooden Clog Village" in utilizing cultural symbols, history, function, the narrativeness, and goods made by designers according to their own unique features gain even more advantages. Nevertheless, the local cultural and creative products of "Caotun Township" enjoy "local uniqueness", because the goods are closely linked with the regional culture and directly affect the local residents in their creation and development of the goods. To analyze and understand the characteristics of local cultural and creative products, we must find out the elements of symbols, history, story, uniqueness and function, which have direct effects on consumers' desire to buy the goods. Local cultural and creative products without these elements can cause cognitive problems of the goods perceived by mass consumers, and reduce their uniqueness (Fig. 3.49).

3.4 Consumers' Perceptions of Cultural and Creative Products

The model derivation of cultural sensibility by Taiwanese scholar Ping-Hong Kuo (2016) explains that the ability for one to identify with a cultural and creative product is essentially the process of sensorial comprehension. Figure 3.50 illustrates the process through which consumers develop an understanding of cultural and creative products. During this process, consumers or tourists are stimulated by external physical stimuli to develop an initial impression of the local cultural industry. Now with a basic impression, consumers are able to go to local industries in person and experience the five senses of sight, hearing, touch, smell and taste. At the same time, when entering the local industry, one can see how the local residents have spontaneously developed their community and explored the unique cultural center of a township. This experience in turn causes tourists to feel a complex set of emotions, including a transition from intuition to comprehension, during which they will go through the key stages of perception, recognition, imagination, and understanding. Tourists' accumulation of local cultural knowledge, along with the historical and cultural experiences of early settlers, results in a process where external physiological intuition is shifted to transition into internal psychological emotions. Thus, through identifying with the historical, storied, symbolic, and signedness of local culture, the distance consumers naturally feel towards local cultural and creative products is

Table 3.9 Creative straw-based merchandise in Caotun (Caotun Straw Crafts Museum, 2005)

Product name	Indoor healthy straw slippers	Straw pot plant	Traditional straw sandals	Straw sandal wall decoration
Cultural and creative products				
Product name	Miniature straw house	Scarecrow pencils	Tea coaster	Straw cards
Cultural and creative products				
Product name	Assorted straw-based animal toys for children	Innovative straw figurines	Straw based knives	Straw Thermos
Cultural and creative products				

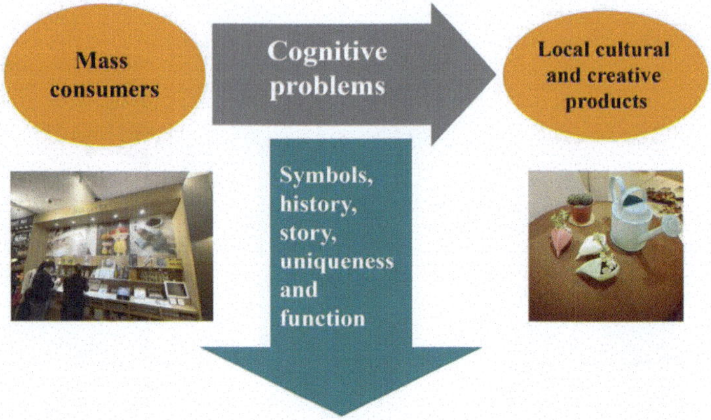

Fig. 3.49 There is a understanding gap between mass consumers and local cultural and creative products

reduced. Through external stimulation and the sensorial experience of local industries, compounded with the process of understanding and eventual emotional attachment, consumers develop a deep understanding of local culture. This process prompts an interest in cultural creative products, which in turn results in the purchasing of said products by the consumer.

The cultural and creative products in Taiwan's local cultural industries call for considerable cognition and understanding of the local cultural knowledge by consumers. The cultural and creative products make the customers generate the feeling of cultural identity, which in turn urges the tourists to consume the goods. The feeling of cultural identity refers to the historical significance, narrativeness, symbolism and symbols, etc. These elements constitute the spiritual connotation of the local cultural and creative products. For people who are engaged in the design and development of the local cultural and creative products, how they interpret the cultural connotation of the symbolism of local symbols, history narrativeness, and unique spiritual meaning become an important issue.

3.5 Relevant Domestic and International Research and Positioning

Local cultural and creative products must be recognized by foreign tourists and local residents alike. In this process of understanding the products through local cultural knowledge, the experience of early settlers and production experience, consumers can comprehend the stories, history, symbols and uniqueness of cultural and creative products. Which in turn results in the desire to purchase these products. Therefore, this book attempts to use semiotic theory to construct development models, master local

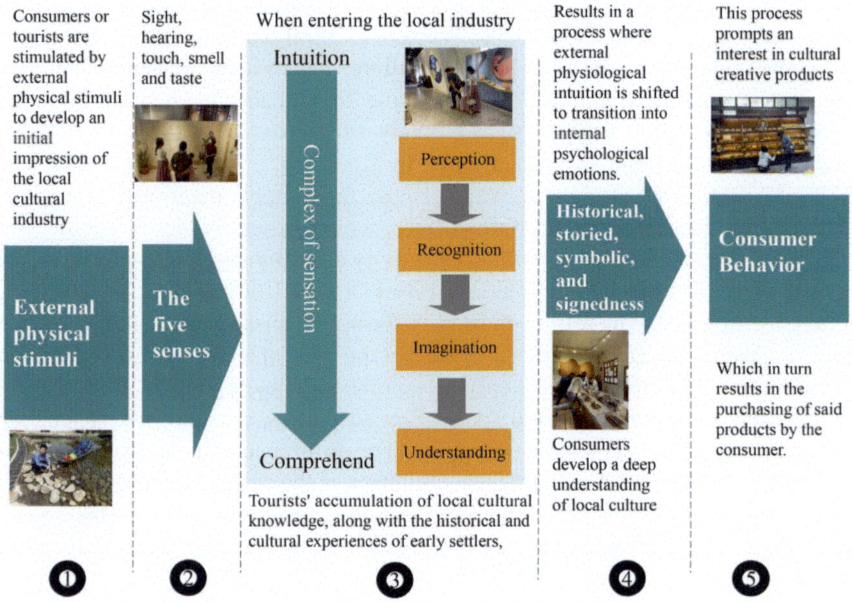

Fig. 3.50 The comprehension process of cultural and creative products

elements, and assist designers in designing products. During this process, we collect related domestic and foreign academic literature on the design and development of cultural and creative products, analyze their respective advantages and disadvantages, and establish the position of this study regarding the direction of cultural and creative product design.

Table 3.10 contains related studies on the design of cultural and creative products at home and abroad. Including this study, there are 12 studies in total. The article entitled "Culture-orientated product design" by Moalosi et al. (2007). The article described about making products become the medium of cultural transmission through design, providing culture-oriented design of culture-based creative products, besides the use function, the designed products can also meet the requirements of consumers in emotion, aesthetics and social contact. The paper discussed about developing relevant aboriginal products by introducing cultural factors analysis, cultural level analysis, design method, cultural product attribute and design evaluation into aboriginal cultural products. The article entitled "A study on the application of an artificial neural algorithm in the color matching of Taiwanese cultural and creative commodities" in Color Research and Application by Hsiao et al. (2014), dividing Taiwanese culture into surface implication, behavior implication and spiritual implication, and decomposing color and vocabulary of culture. The article entitled "Culture-Based Knowledge Towards New Design Thinking and Practice A Dialogue" by Leong and Clark (2003), explaining that the design elements of cultural products cover the concept of reverse-triangular cultural space of tangible things, behaviors and internal layer, introducing Eastern and Western cultural features, and design pattern of the convergency of Eastern and Western cultures in two reverse-triangular cultural space. The journal paper entitled "Transforming Taiwan aboriginal cultural features into a modern product design: A case study of a cross-cultural product design model" in International Journal of Design by Lin (2007). The book entitled "Cultural Foundation of the Design" by Yang (1998). In the book, mentioned that culture is to distinguish and understand human lives, which generally can be divided into three layers: metaphysical layer (thinking activity and language), mid-physical layer (social and interpersonal interactions) and physical layer (tools used by human beings and tangible objects). The article entitled "Cultural Realism and Virtualism Design Model" by Wang (2020), The design model for the present study is based on national and international literature that analyzed symbolism, metaphysics, three layered culture, reverse-triangular cultural space, Zen aesthetics, etc. and honest feedback based on the actual feeling toward the culture. The difference of Laozi's philosophy in this model can contribute to the different cognition of the culture in various aspects in the time axis and provide diverse views of the culture. Therefore, this study proposed the theory of "Cultural Realism and Virtualism" by introducing philosophical and mathematical methods and applying the concept of "virtualism" and "realism" to deconstruct and construct cultural elements and implement the design of cultural and creative products such as complex variable, fuzzy theory, structure variation and color evaluation. The article entitled "A Study on Fuzzy C-means Application in Austronesian Language Cultural and Creative Product Colorss" in Color Research and Application by Wang (2018), providing the application of Fuzzy C-means on cultural and creative products for designers. Through the result of the design of case, cultural and creative products developed from Taiwan and New Zealand, fulfilled the purpose that the theory has been realized in practice, and the products successfully entered the markets. Among them, Lee (2004), in The Application to The Symbol of Cultural Commodity—With a Case Study of Cultural Creative Industry in Hu-ban village, proposed semiotics as a study of theoretical design, and Ho et al. (1996) in Some Speculations on Developing Cultural Commodities, developed cultural and creative products through the four levels of style leadership, market strategy, design expression and shape elements. In addition, in Zheng's (2004) paper titled The Application of Taiwanese Aboriginal Culture in Product Design, cultural and creative products were designed by breaking culture down into three separate levels: the first layer consists of tangible forms, the second layer consists of behavior, and the third layer consists of ideological patterns. At the same time, Su's (2004) paper, The Design added-value Model for Aborgines' Culture Merchandise- Amis Crafts as examples, uses factor elements to design products. The last study, Chen's (2005) Study On the Cooperative and creative design works for local culture commodities–with a case study of Caotun straw goods, which details the cooperative effort between academia and local product design departments to reach a consensus on the development of cultural and creative products.

To summarize the views detailed above, the following are a few general observations; first, if the local residents do not have professional design training, they are

Table 3.10 Domestic and foreign research on creative product design

Researcher	Topic	Time	Source	Method
Moalosi et al. (2007)	Culture-orientated product design	2007	International Journal of Technology and Design Education	Making products become the medium of cultural transmission through design, besides the use function, the designed products can also meet the requirements of consumers in emotion, aesthetics and social contact
Hsiao et al. (2014)	A study on the application of an artificial neural algorithm in the color matching of Taiwanese cultural and creative commodities	2014	Color Research and Application	Surface implication, behavior implication and spiritual implication
Wang (2018)	A study on fuzzy C-means application in Austronesian language cultural and creative product colors	2018	Color Research and Application	Providing the application of Fuzzy C-means on cultural and creative products for designers
Wang (2018)	Cultural realism and virtualism design model	2020	Springer	This study proposed the theory of "Cultural Realism and Virtualism" by introducing philosophical and mathematical methods and applying the concept of "virtualism" and "realism" to deconstruct and construct cultural elements and implement the design of cultural and creative products such as complex variable, fuzzy theory, structure variation and color evaluation

(continued)

Table 3.10 (continued)

Researcher	Topic	Time	Source	Method
Leong and Clark (2003)	Culture-based knowledge towards new design thinking and practice-A dialogue	2003	Design Issues	Convergency of Eastern and Western cultures in reverse-triangular cultural space
Lin (2007)	Transforming Taiwan aboriginal cultural features into modern product design: a case study of a cross-cultural product design model	2017	International Journal of Design	Factors analysis, cultural level analysis, design method, cultural product attribute and design evaluation
Yang (1998)	Cultural foundation of design	1998	Ai-Tai Publishing	Metaphysical layer (thinking activity and language), mid-physical layer (social and interpersonal interactions) and physical layer (tools used by human beings and tangible objects)
Ho et al. (1996)	Some speculations on developing cultural commodities	1996	The Journal of Design	This study reconsidered the communication of design message through the encoding process of developing a product. Some preliminary concepts have been explored by examining typical and somehow more successful cases. As a result, a couple of ideas are proposed for the future design practice and further in-depth design research

(continued)

Table 3.10 (continued)

Researcher	Topic	Time	Source	Method
Lee (2004)	The application to the symbol of cultural commodity—with a case study of cultural creative industry in Hu-ban village	2004	Department of Industrial Design, National Yunlin University of Science and Technology	How cultural symbol are attached into commodities by translating symbolic codes is an interesting issue and worthy of being discussed. However,it is necessary for us to consider that how cultural symbols are choosed made-up,translated,attached and merged to commodities when culture are translated to commodities. Therefore,this praject will focus on the present when cultural symbols are translated to commodity
Chen (2005)	A study on the application of an artificial neural algorithm in the color matching of Taiwanese cultural and creative commodities	2005	Department of Industrial Design, National Yunlin University of Science and Technology	There are three conclusions in this research: First, depend on participant who is designed together and experience of communication processes, order out the cooperative mechanism of designed together. Second, from the processes of design break-in and intention inspire, analyzing the procedure of designed together.Third, come out with the possible way that is use the local culture symbol in the common design and intention inspire

(continued)

Table 3.10 (continued)

Researcher	Topic	Time	Source	Method
Su (2005)	The design added-value Model for Aborgines' culture merchandise- amis crafts as examples	2004	Department of Industrial Design, National Cheng Kung University	The suggestion factors needed to be first considered in each stage of the model are stated for reference when designing culture products. And they are: for "culture topic selection," the first consideration factor is the cultural totem, followed by ancestors' tales and legends and characteristics of life behavior; for "framed conditions of products," the factors are product attributes, market segmentation and product price; for "design presentation," the factors are colors, texture and materials; for "vending spots," the factors in order are participating in exhibitions of related topics in a mobile way, selling online, diversified operation and selling at sight-seeing spots; for "personal selling," the factors in order are owners' cultural traits and narration, agents, tour guides and loyal customers; for "ways of selling," they are monopolization by sellers, consignments from artists, and self-selling by artists; as for sales promotion," the factors considered are selling in the culture season, giveaways, alliance with different industries, and exhibition design

(continued)

Table 3.10 (continued)

Researcher	Topic	Time	Source	Method
Zheng (2004)	The application of Taiwanese aboriginal culture in product design	2004	Department of Industrial Design, Chang Gung University	The research products include ceremonial drinking cup, pottery container, naked sculpture and wooden boat such traditional implements of Taiwan aborigine. First, we sort the data and use digital collection approaches to be references for consequential product designs. In the second phase, we establish the learning model of Taiwan aboriginal culture through creative learning, milieu story and product semantics such approaches. The third phase is value added to cultural product design creation, that is, process practical design for cultural products by using preceding creative leaning and cultural product design model; through practical operation, we use works handed over by attending trainee to process discussion and design result analysis. Finally, we employ multidimensional measurement to design evaluate. In addition to verify whether design results origin from the stimulation of traditional implement of aborigine and seek for the preference ratings within consumers' perception and applicable creation attributes

unable to rely on the product development goals outlined in the theoretical frameworks of academia in the long-run. Second, academia is heavily focused on theoretical frameworks, yet these frameworks are unable to achieve the goals set when it comes to product development. Third, although it is possible to conduct experiments, studies, and other forms of quantitative research with relevant local designers, without long-term academic or professional design guidance, relevant users are unable to become self-reliant and succeed in short-term product development. This study aims to take a step forward in examining and improving the three issues mentioned above. This book, On the Creative Product Design of the Cultural Codes, divides the theories of various semiotic scholars into two levels: one is cultural extension meaning, which refers to shape, color, material, texture and function; the other is cultural connotation meaning, which refers to cultural stories, legends, etc. The four advantages we aim to gain from such a division are 1. Simplify the process of looking up data; 2. Shorten the development time of designers; 3. Aid in foreign developers' ability to master local cultural resources; 4. Aid in local designers' mastery of related elements. With respect to the symbolic, historical, and unique spiritual aspects of local symbols, this book raises two points: one being "helping designers in the development of cultural and creative products" and the second "the differences in understanding regarding cultural and creative products among mass consumers", forming the direction of the research conducted in this book (Fig. 3.51).

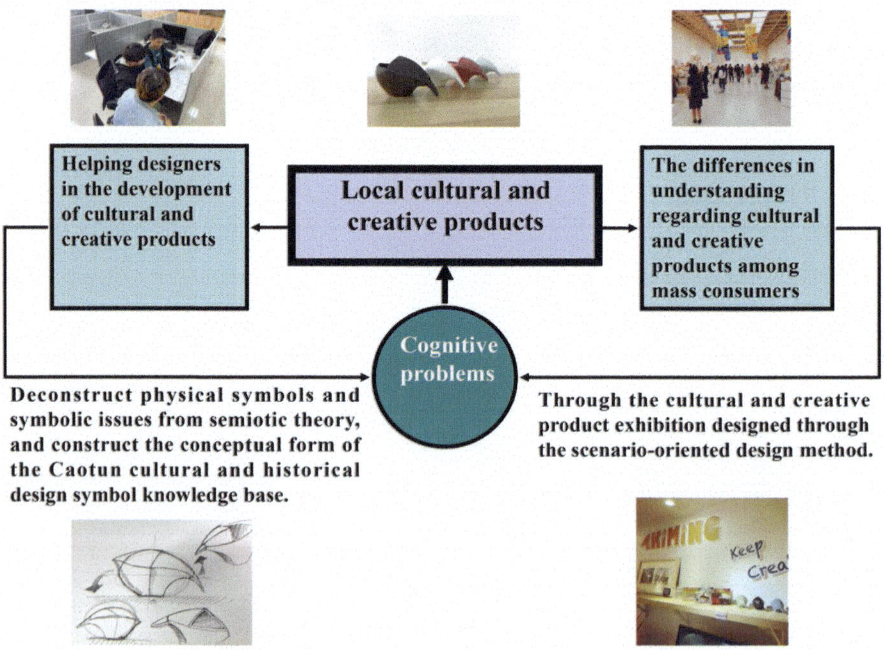

Fig. 3.51 Solving the comprehension issue among designers and consumers

1. "Helping designers develop cultural and creative products": Deconstruct physical symbols and symbolic issues from semiotic theory, and construct the conceptual form of the Caotun cultural and historical design symbol knowledge base. Assist local residents in its implementation during the planning of community development operations. Provide a knowledge base of local culture for foreign and local designers to aid in the design of local cultural and creative products.
2. "The differences in understanding regarding cultural and creative products among mass consumers": Through the cultural and creative product exhibition designed through the scenario-oriented design method, strengthen the storytelling aspect of products, and bridge the knowledge gap between foreign tourists, local residents, and local cultural and creative products.

References

Caotun Straw Crafts Museum. (2005). *Creative straw-based merchandise in Caotun*. Retrieved from https://sixstar.moc.gov.tw/blog/ttshoe/communityAction.do?method=doCommunityView

Chen, C. N. (1995a). *Overall community creation and cultural industry development* (pp. 4–9). National Taiwan Craft Research Institute.

Chen, C. N. (1995b). *Culture, industry symposium and overall community building China-Japan exchange exhibition proceedings*. Council for Cultural Affairs, Executive Yuan.

Chen, Z. S. (2002a). The contribution of ancient documents to the historical research of Caotun area. *Feng Chia Journal of Humanities and Social Sciences, 5*, 107–126.

Chen, C. L. (2002b). *Analyze the cultural production and media role of the economic market-four cases as an example*. Master's thesis. Department of Communications Management, Shih-Hsin University, Taipei.

Chen, P. (2005). *Study on the cooperative and creative design works for local culture commodities–with a case study of Caotun straw goods straw goods*. Master's thesis. Department of Industrial Design, National Yunlin University of Science and Technology, Yunlin.

Compiled by Caotun Town Office. (1986). *Caotun Township Chronicle*. The Caotun Township Office.

Compiled by Caotun Town Office (2005). *Continued Edit of the Caotun Town Chronicle*. The Caotun Township Office.

Council for Cultural Affairs, Executive Yuan. (1999). *Concepts and examples of overall community development* (2nd ed.). Taipei: Council for Cultural Affairs, Executive Yuan.

Feng, J. L. (2002). *Culture is good business*. Chengbang Publishing House.

Grandpa Wen Caitang. (2022). *The story of the magnolia coco*. Tainan City Government.

Ho, M. C., Lin, C. H., & Liu, Y. C. (1996). Some speculations on developing cultural commodities. *The Journal of Design, 1*(1), 1–14.

Hsiao, S. W., Wang, M. F., Lee, D. J., & Chen, C. W. (2014). A study on the application of an artificial neural algorithm in the color matching of Taiwanese cultural and creative commodities. *Color Research & Application, 40*(4), 341–351.

Huang, F. X. (2001). *Discussion on the Spontaneity of local residents and the revitalization of local traditional crafts—using Yingge ceramics as an example*. Master's Thesis, Department of Industrial Design, Tunghai University, Taichung.

Hwang, S. H. (1998). On Japan's Regional Cultural Industry Promotion Policy. *Interior Space Magazine*, 106–107.

Hwang, S. H. (2001). Cultural Industry and Residents' Participation. In *Proceedings of the General Meeting of Community Development of the Council for Cultural Affairs Held in the 90th year of the Republic of China* (pp. 19–28). Taipei: Council for Cultural Affairs, Executive Yuan.
Kuo, P. H. (2016). Effects of synchronous web-based instruction on students' thinking styles and creativity. *Eurasia Journal of Mathematics, Science & Technology Education, 12*(3), 609–619.
Li, C., Peng, R., Gu H., Peng Q., & Wang, J. (2008). Magnolia Coco-Hakka Yuanxiang, Council for Hakka Affairs, Kaohsiung City Government.
Lee, Y. F. (2004). *The application to the symbol of cultural commodity –with a case study of cultural creative industry in Hu-ban village.* Master's thesis, Department of Industrial Design, National Yunlin University of Science and Technology, Yunlin.
Leong, B. D., & Clark, H. (2003). Culture-based knowledge towards new design thinking and practice-A dialogue. *Design Issues, 19*(3), 48–58.
Lin, R. T. (2007). Transforming Taiwan aboriginal cultural features into modern product design: a case study of a cross-cultural product design model. *International Journal of Design, 1*(2), 45–53.
Lin, G. F. (1998). *The 1998 National Art Season, a Special Issue on the Results of the Local Cultural Festival Planning Camp* (pp. 62–68). Taipei: Council for Cultural Affairs, Executive Yuan.
Lull, J. (1995). *Media, communication, culture: a global approach.* Columbia University Press.
Moalosi, R., Popovic, V., & Anne, H. H. (2007). Culture-orientated product design. *International Journal of Technology and Design Education, 20*(2), 175–190.
National Culture and Arts Foundation. (2004). *A book on the practice of cultural and creative industries.* Taipei: Business Weekly.
National Arts and Cultural Activities Information System of the Ministry of Culture. (2022). *Arts and cultural activities information system,* Retrieved from: https://event.moc.gov.tw/mp.asp?mp=1.
Su, M. L. (2005). *The design added-value model for Aborgines' culture merchandise-Amis crafts as examples.* Master's thesis, Department of Industrial Design, National Cheng Kung University, Tainan.
Taiwan Design Research Institute. (2005). *DESIGN creativity-design-brand marketing,* October and November bimonthly. Taiwan Design Research Institute.
The National Palace Museum Shop. (2022). *National palace museum shop.* Retrieved from https://www.npmshops.com/.
The National Palace Museum. (2022). *National palace museum.* Retrieved from http://www.npm.gov.tw/.
Throsby, D. (2001). *Economics and culture.* Cambridge University Press.
United National Educational, Scientific and Cultural Organization. (2014). Cultural industries: a challenge for the future of culture, Retrieved from http://www.unesco.org/ulis/cgi-bin/ulis.pl?catno=49972&set=00548AFF7B_3_30&gp=1&lin=1&ll=1.
Wang, L. J. (2003). *Economic discourse in cultural politics: from elite culture to mass economy?* (pp. 1–18). Annual Meeting of the Society for Cultural Studies.
Wang, L. J. (2005). Economic discourse in cultural policy: from elite culture to cultural economy. *A Journal of Cultural Studies, 1,* 169–195. https://doi.org/10.6752/JCS.200509_(1).0007
Wang, M. F. (2018). A study on Fuzzy C-means application in Austronesian language cultural and creative product colors. *Color Research and Application, 43*(3), 375–386.
Wang, M. F. (2020). *Cultural realism and virtualism design model.* Springer.
Wang, M. F. (2022). Proposing a new oceanic culture-based creative product design method to construct and deconstruct products of the Austronesian aborigines (SCI). *Journal of Coastal Research, 38*(2), 458–470. https://doi.org/10.2112/JCOASTRES-D-21-00045.1
Wang, M. F., & Cheng, C. F. (2021). Hakka Faith Sanshan King Cultural and Creative Product Design and Research, *National Science Forum, Pingtung: National Museum of Marine Biology and Aquarium,* 229–232.
Wang, M. F. (2007). *The research of symbolism in local culture commodity developing mode—a case study in Nantou county.* Master's Thesis, Department of Industrial Design, Tunghai University, Taichung.

Weng, X. D. (1995). Local development and territorial revitalization. *National Taiwan Craft Research Institute, 55*, 10–15.
Wooden Clog Museum. (2022). *Baimi Wooden Clog Village*. Retrieved from https://sites.google.com/view/baimivillage/home-%E4%B8%BB%E9%A0%81
Yang, Y. F. (1998). *Cultural foundation of design*. Ai-Tai Publishing Co.
Yilan Baimi Clogs Village. (2022). *Baimi Wooden Clog Village*. Retrieved from https://sites.google.com/view/baimivillage/home-%E4%B8%BB%E9%A0%81.
Zheng, H. (2004). *The application of Taiwanese aboriginal culture in product design*. Master's thesis, Department of Industrial Design, Chang Gung University, Taipei.

Chapter 4
Construction of the Knowledge Base Concept Form and Process

The book collected the literature related to the theory of semiotics and found that the theory could interpret the local cultural symbols. We extracted the symbol elements, and proposed to construct the knoewledge base of cultural and historical design symbols to shorten the designers' gap of cultural cognition difference, and help to develop the local cultural and creative products. Then, to check the feasibility of the knowledge base of cultural and creative products design and development model, we created some goods from our experience to test the feasibility of the goods development model.

4.1 Literature on Semiotics

Baudrillard said that "In order to become an object of consumption, the object must become a sign." In the discussion of "symbols," most people think that "symbols" are a concrete crystallization of human thought and experience. It covers a wide range of content, as long as it is a visible form related to human spiritual culture, such as language, writing, religion, art, thought, an object can be regarded as the product of the symbolic thinking mode. This is because these visible forms have the characteristics of description, meaning, imagery, feeling and so on that the symbol itself has. Commodities must be symbolized so that consumers can develop a stronger understanding of its existence. The fact that commodities are "useful things" is taken for granted, and at the same time, consumers take the attached "meaning" behind each product as the criterion for selection and purchase (Hoshino, 1991). Commodities cannot but be turned into "symbols" that appeal to consumers' sensibilities. Designer-branded merchandise can act as a "symbol of differentiation" for the owner. Being the product of a designer is at its core, and the brand structure is the same as what anthropology calls a "totem." Once a commodity is established as a brand, it transcends its physical characteristics and has a certain symbolism, representing the "totemization" of the product. Since commodities have become symbols of social value, they can be

M.-F. Wang, *Creative Product Design With Cultural Codes*,
https://doi.org/10.1007/978-981-19-6767-2_4

used as symbols of social class distinction (Mirzoeff, 1999). Therefore, the cultural creative products of Caotun are not limited by the pure utility of the commodities themselves, but also have the added symbolic meaning of local cultural symbols.

4.1.1 Relevant Theories of Various Semiotic Scholars

This section aims to summarize the common characteristics of symbolic interpretation from the theories of various semiotic scholars such as Charles Peirce, Ferdinand de Saussure, Charles Morris, Roland Barthes, Umberto Eco and Hoshino Katsumi, in order to build the necessary basic concepts for the construction of the cultural creative product development model.

1. Charles Peirce's semiotic theory

Peirce categorizes symbols into three types: Icon, Index, and Symbol, to interpret the various relationships between "Signature" and "Meaning."

(1) Icon: an icon represents the similarity between the indexed thing and its subject, and the relationship between any two indexed objects are not random and arbitrary. Just like a person will appear similar no matter if they are perceived as in a picture or seen in person. Looking at a picture of a person is just like seeing the person in real life. This image also represents the symbol of this person. For example, the straw sandals of Caotun township.
(2) Index: a symbol used to judge the relationship between a linking object and its subject by means of an actual linking object, meaning, there is an inherent causal relationship between the symbol and the referent. Just like when seeing smoke means that there is a fire, sneezing means one has come down with a cold. Just like a walking porter.
(3) Symbol: the relationship between the "signifier" and "signified" of the symbol itself is arbitrary and conventional. For example, peony flowers in China represent a symbol of wealth, and shaking hands with people expresses friendly greetings. This phenomenon produces different cognitive structures through differences in social backgrounds and cultural structures (Berger, 1933; Fiske, 1989). Example: Interpreting the story of the Caotun Lugang Danpushe (Table 4.1).

Peirce gave symbols more space, believing that a symbol can be explored from three angles, and divided into icons, meaning that the straw sandals seen in the photo are very similar to the shoes worn by ancestors, this image means that the straw sandals can be witnessed. The index refers to when you see people walking in the rice fields, you can think of farmers. Symbols refer to words that have a traditionally agreed upon meaning. For example, when you open a history book and see the local word "Lugang Danpushe," it represents the symbolic story of Caotun.

Table 4.1 Interpreting Caotun's three types of cultural codes

Three types of symbols in Caotun	 Icon	 Index	 Symbol
Characteristics	Can be witnessed Ex: straw sandals can be witnessed	Can be imagined Ex: the connectivity of the porter	Must be learned Ex: learning history through "Lugang Danpushe"
Method of expression	Image similarity Ex: straw sandals are made of straw	Logical similarity Ex: a porter walks along his route	Tradition, convention Ex: the tradition of writing in Caotun
Examples	Straw sandals photo Ex: photographs of straw sandals	Farmland, hats, poles Ex: porter	Words, numbers Ex: history books

2. Ferdinand de Saussure's semiotic theory

The Swiss linguist Saussure believed that a sign is an entity with meaning; a sign is composed of a signifier and the signified. As far as language symbols are concerned, the symbol is a Sound-Image; while the symbolic meaning is a concept that the symbol refers to (Su et al., 2007). The following section defines the signifier and the signified.

(1) Definition of Signifier and Signified

A. Signifier: Symbolic forms or specific images that appear as objects, which can be obtained from the senses, such as the audiovisual effects of sounds, the visual image of words, the color, shape, or pattern of objects (Berger, 1992). For example, the straw sandals in Caotun are an important product of local cultural industries and have become a symbol of local objects.
B. Signified: The meaning behind the symbol that appears as an object, the imagination or association of the object, is an abstract emotion, belief, and concept; it can also be determined according to the overall social and cultural customs, and it is a psychological concept (Berger, 1992). The meaning behind the straw sandals in the locality is the Lukang Danpushe, which has served as an important medium for cultural identification, and promoted the recognition and maintenance of the environment (Fig. 4.1).

(2) Paradigmatic and Syntagmatic

The paradigmatic and syntagmatic are part of a group of ideas derived from Saussure. It is equivalent to the metaphor and metonymic used by Jakobson. In fact, the former, paradigm and metaphor would mean commutation, and the latter, syntagm and metonymy refers to a permutation. Therefore, in the modern day, a good way to

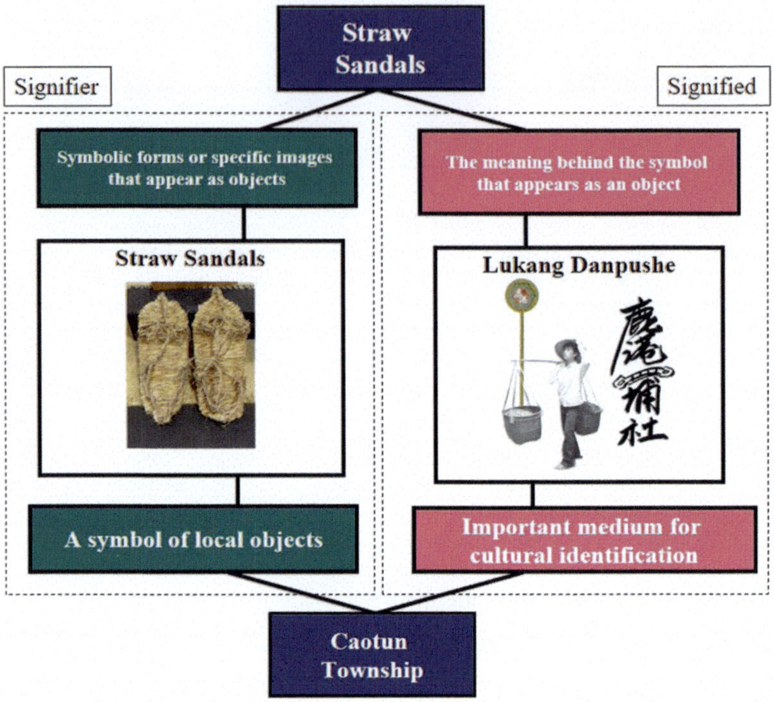

Fig. 4.1 Caotun signifier and Signified

describe the relationship between the paradigm and the syntagm are as follows. The paradigm is like a family tree, as it develops linearly, while the syntagm develops in a collateral series (Ng, 2018). Take shoes as an example: straw sandals, cloth shoes, leather shoes, and slippers can be called paradigms; while the syntagm of shoes can be regarded as sportswear, sports hats, and other peripherals to go along with sneakers or shoes.

Saussure's theory of paradigm and syntagm presents an argument regarding what a designer calls "lateral and vertical thinking." For example, the Caotun Township Chronicle records that during the Qianlong and Jiaqing eras, the Lugang Danpushe passed by Cao Xie Dun (Caotun) (Compiled by Caotun Town Office, 1986), and wanted to rest, drink tea, and replace their shoes during their stay; these actions can be considered syntagmatic (similar to lateral thinking). Replacing their shoes can be considered to be a paradigmatic action (similar to vertical thinking), along with actions such as wearing out their shoes, buying new straw shoes (sandals), putting them on, and continuing on their journey. See Table 4.2.

3. The semiotic theory of Charles Morris

Morris's syntax expresses the formal relationship of a mark and its relationship with other marks. Semantics, which represents the relationship between tokens and

Table 4.2 Syntagmatic and paradigmatic analyses of Caotun's culture

Syntagmatic (lateral thinking)	Paradigmatic (vertical thinking)			
	Paradigmatic 01	Paradigmatic 02	Paradigmatic 03	Paradigmatic 04
Syntagmatic 01	Wanting to rest	Under a tree	Area with shade	Rest one's eyes
Syntagmatic 02	Wanting to drink tea	Tea shop	Looking for a table	Rehydrate with tea
Syntagmatic 03	Wanting to exchange shoes	Worn out	Buy new straw shoes (sandals)	Continue on the journey

objects and their meanings. Pragmatics, which expresses the relationship between the sign and the user of the sign, that is, the interpreter, has become a widely used subject classification. Among them, the proposal of "pragmatics" is the most distinctive, because in the field of semiotics, scholars study the process of using signs or the relationship between people and the signs used. Then pragmatics involves the psychology of the speaker, the response of the audience, along with the social and linguistic context of the discourse. Much of the classification and object of discourse is related to the process of language application. "Syntax" is primarily concerned with the formal properties of language structures in relation to the symbols themselves. "Pragmatics" is mainly concerned with the relationship between linguistic signs and extralinguistic referents, and the relationship between signs and the things they describe (Yu, 2001). The relationship between the three is shown in Fig. 4.2.

4. Roland Barthes' semiotics

The core of Roland Barthes' theory is about identifying that symbols contain two levels of meaning, the important elements of which are the two levels of denotation and connotation mentioned in Fig. 4.3.

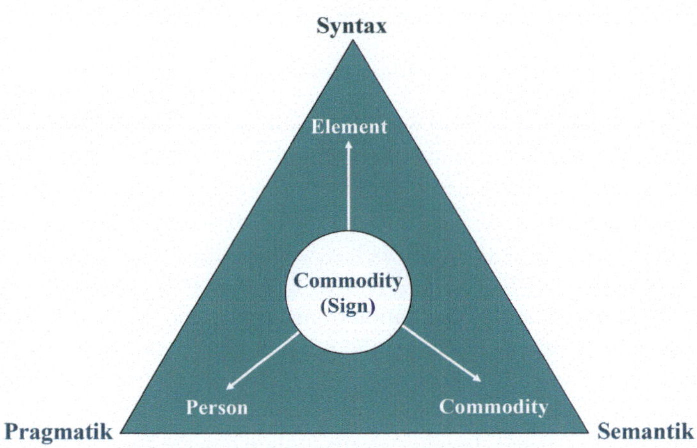

Fig. 4.2 Charles Morris' three properties of symbols

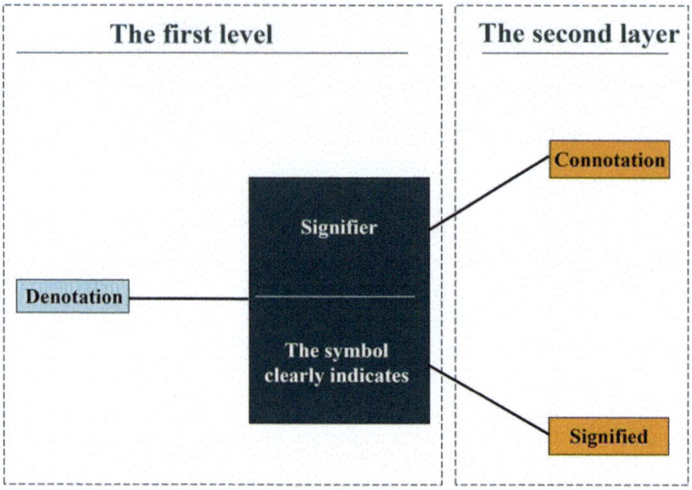

Fig. 4.3 Roland Barthes' definitions of denotation and connotation

(1) The first level of meaning is denotation: that is, the meaning that the symbol clearly indicates and belongs to the category of general common sense (Chang et al., 2009).
(2) The second layer of meaning is connotation: It can be viewed in three different ways. Connotation refers to the subjective or relatively subjective feelings constructed in the user's mind through the use of symbols. Of course, this feeling or relationship interacts with people's cultural values, so the interpreter itself and the object (referent) represented by the symbol will affect the interpretation of the symbol (the signified). For Barthes, the first layer of signifiers is the symbol of connotation, and according to this first level of signifiers, people look for the connotation within the second layer (Chang et al., 2009).

5. The semiotic theory of Italian scholar Umberto Eco

Umberto Eco proposed two major constructions from the various theories regarding linguistic symbols, namely "grammar theory" and "sign making theory." Eco was, on a daily basis, devoted to the research of literary semiotics, aesthetics, epistemology, semiotics, and structuralism. Eco continues Peirce's direction of thinking and studies the communication process in it. He believes that with the help of semiotics, all cultural events can be analyzed. He believes that the study of semiotics exists as long as the communication of information transmission can function on the basis of codes. Eco believes that if the code of the product is not introduced to the society through learning, the so-called creed of functionalism is just form and function, and it is still only a fantasy. If the code cannot be explained, it is impossible to manipulate various products (Eco, 1984). In other words, only when the form of the appearance can clearly indicate the function, can the use of the product be realized smoothly. Eco also applies important concepts in design: Denotation and Connotation; Eco interprets denotation as the message received by the recipient of an expression in a culture, Conversely, connotation is understood as all things that remind the recipient of the meaning of a sign; a connotation can be seen as a collection of associations based on a certain sign in a particular culture. In terms of design, connotation refers to things which can be used to understand the message that the product represents. The purpose of the mark in the product is to guide the proper execution of the action, and the form that the product exhibits is based on commonly agreed upon expectations and a sociocultural system, so that the product's ultimate purpose can be achieved. According to Eco's research, the same sign may express different meanings when present in different cultural contexts (Lee, 2002).

(1) Denotation: this is limited to what can be referred to, and for others, it is basically similar. In the language of signs, representing the result of passive use of signs, the highest goal is the practical function; on a certain level, the "structural" function is conveyed (Yu, 2001). For example, Dengying Academy in Caotun Town is a national third-level historic site.
(2) Meaning: It is emotional. Everyone feels or expresses it with a certain value or preference. Different people have different feelings about the same symbol, so it will also cause misunderstandings in communication. In the language of signs, it represents the result of active use of signs, and the highest goal is aesthetic function; on a certain level, it reflects different "symbolic meanings (Yu, 2001)." For example, the connotation of Dengying Academy in Caotun Town symbolizes the local education school dating from the Qing Dynasty. Due to the changes of the times, it has now become a temple dedicated to Emperor Wenchang. It is a place where local people pray for high academic achievement and strong academic progress (Table 4.3).

6. Hoshino Katsumi's semiotic theory

Hoshino Katsumi proposed that the value of commodities is that commodities have meaning = "denotation." "Value in use," according to the previous common sense, is a clearly visible meaning on the surface of a commodity. "Symbolic value," on the other hand, is the hidden meaning in commodities that were previously regarded

Table 4.3 Meaning and denotation of Caotun's Dengying academy

Symbols of the landscape	The third level	
	Sa: Dengying academy is dedicated to Emperor Wenchang	Se: high academic achievements, strong academic progress (connotation meaning)
	The second level	
	Sa: religious celebrations at the Dengying Academy	Se: spring and autumn festival activities
	The first level	
	Sa: Dengying academy	Se: appearance of the third-level historic site (extension meaning)

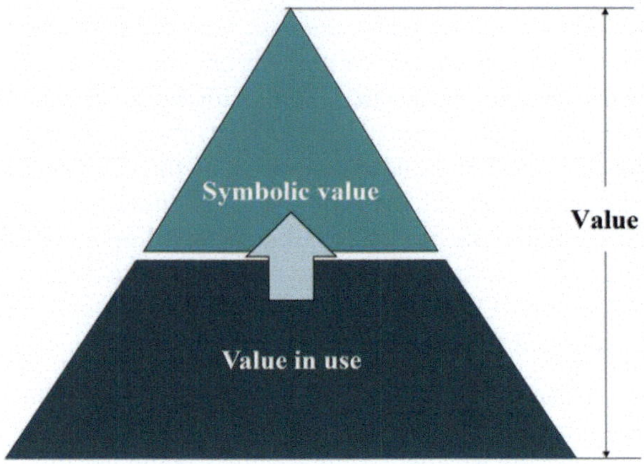

Fig. 4.4 Commodity value (Hoshino, 1991)

as invaluable. Such as: the sweet taste and alcohol content of beer, called "value in use." From that, the design of the container and the surrounding parts that make a fizzing sound when the beer is poured can be called "symbolic value." The distortion of commodities, based on the conversion of the above-mentioned values, can be said to have prompted a 180-degree change in people's understanding of commodities (Hoshino, 1991). Expressed by straw health slippers, a local cultural creative product in Caotun, the transition of the product went from its cultural "value in use" to its cultural "symbolic value (Fig. 4.4)."

4.1.2 Symbol Conversion and Interpretation

The above literature puts forward the views of six semiotic scholars on symbols, which gives this book a concept of symbols, as well as ideas on how to convert symbols and interpret symbols. The following will discuss the structure and composition of symbols, the metaphor of symbols as well as how to convert symbols.

1. Symbolic structure and deconstruction

Deconstruction has two meanings: one is to explain the structure, and the other is to break down the old structure and create a new one. The process of dismantling the old structure and establishing the new one can also be regarded as transformation, that is, dismantling the old symbol system and successfully converting all resources into a new symbol system. In the process of disintegrating the old sign system and establishing a new sign system, the two functions of accommodation and assimilation play a very important role. Accommodation refers to the change of the mind at any stage in response to a new environment in order to accommodate new or particular environmental characteristics. On the contrary, depending on the degree of development of the mind's understanding of the new environment, the new environment can be included in a certain level of understanding. This operation is called assimilation

Table 4.4 Accommodation of assimilation and dissimilation among creative straw-based products

Assimilation	Accommodation
Straw animal shaped goods	Indoor healthy straw slippers

The straw animal-shaped products are placed in the museum. We will see it as a display and ornamental function. If it is placed in a primary school classroom, children will play as straw animals, thus displaying the process of assimilation	The Caotun cultural industry has improved the traditional straw sandals over time to become modern slippers, but the product still retains the local characteristics of traditional straw sandals, which is in line with our cognition of traditional straw sandals and displays the process of accommodation

(Piaget, 1970). Table 4.4, immediately following this section, uses the two examples of straw play toys and indoor healthy straw sandals as examples of assimilation and accommodation.

2. Symbolic metaphor

Metaphor comes from the Greek word metaphorein (transformation), which means that the meaning of a word can be related to another word through certain methods, such as association, similarity, and adjacency (Fiske, 2010). A relationship between two words is developed through these methods, and assists in peoples' understanding of each word. So, a metaphor can be said to be a simile, and when we use this metaphor to talk about one thing, we can clearly imply another (Sallie, 1982). For example, police usually drive "black and white cars" to arrest fugitives, and black and white cars refer to police cars. This is because the color of a police car is typically black and white, so when a car is black and white, it will often be associated with a police car, so we will replace the term police car with the term black and white car. So, when metaphors are used, the real meaning is not the literal meaning, but the meaning behind the text. In addition, using the language of everyday life as an example of metaphorical words, "you have the heart of a beast," of course, has nothing to do with animal organ transplant, just as much as the statement "you have the heart of a saint" is not related to being possessed by the metaphysical (Yu, 2001).

3. Saccharin and sandals symbol conversion

Table 4.5 shows the examples of saccharin and straw sandals. It can be found that saccharin and straw sandals have not undergone any changes, saccharin is still saccharin; straw sandals are also still straw sandals; yet people have developed

Table 4.5 Code conversion of saccharin and straw sandals (Birkhoff, 1973; Hu, 2010; Wang, 2007; Yang, 1998)

Product name	Positive or negative perception	Conversion example
f (saccharin)	Negative	Increased digestive burden, artificial, chemical → carcinogenic → death
f (saccharin)	Positive	Eliminate calories, healthy blood vessels → good health
f (straw sandals)	Negative	Declining, uncomfortable, impatient → funeral
f (straw sandals)	Positive	Breathable, environmentally friendly, reminiscent → healthy
f(x) = k, x is the symbol, k = + or k = −, then it depends on the f function		

positive or negative views towards these two items. This is because people regard saccharin and straw sandals as symbols. When these two symbols are placed in different contexts, people will develop different levels of understanding regarding these products after the conversion of the code in each symbol's respective context (Wang, 2007; Yang, 1998).

Based on the above, the concepts discussed by various semiotic scholars, as well as how symbols are deconstructed and structured, are undoubtedly necessary to understand how symbols can grasp local resources and convey design elements through the conceptual form of cultural and historical design symbols knowledge base. And through this conceptual form of the cultural and historical design symbols knowledge base, we further understand how these ideas become important in literature used in the cultural creative product development model. The core of this book is how to use semiotic theory to construct a knowledge base. For details, please refer to the construction content and steps of the cultural creative product development model below.

4.2 Design Models and Procedures of Cultural and Creative Products

See Fig. 4.5.

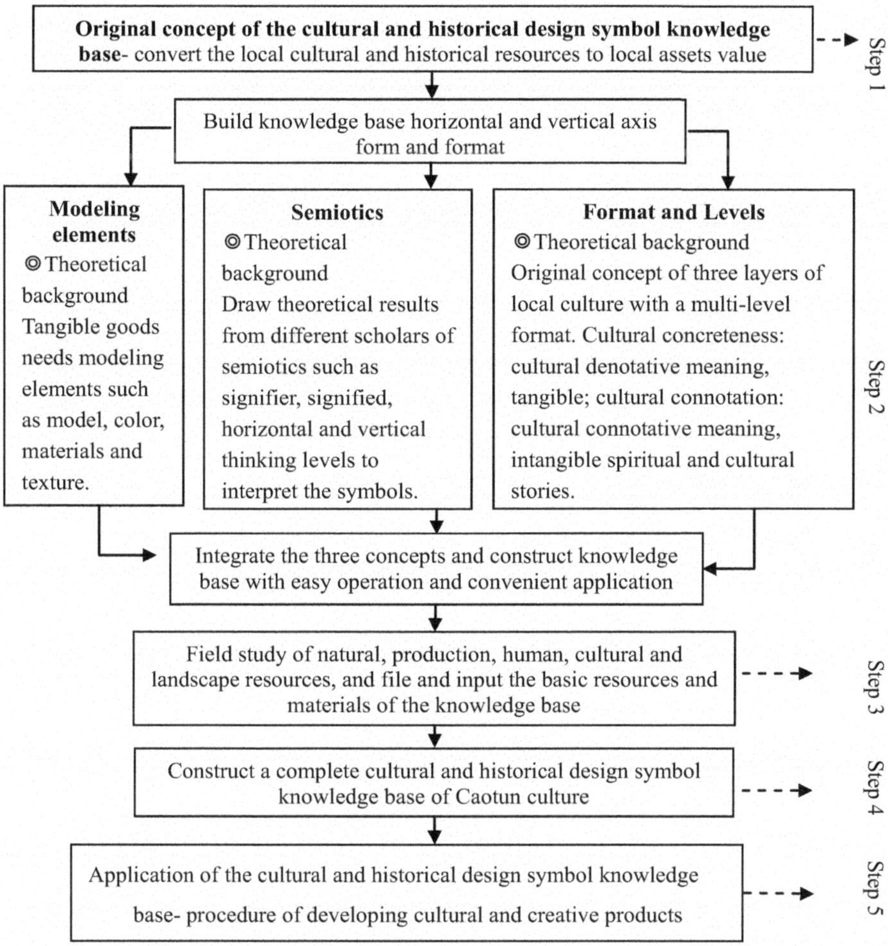

Fig. 4.5 The contents and steps for the construction of cultural and creative products design and development model

4.3 The Contents and Steps for the Construction of Cultural and Creative Products Design and Development Model

Step I: Start from the original concept of knowledge base, understand that local cultural and historical data can create local cultural and creative products value.

Step II: Build knowledge base horizontal and vertical axis form and format, taking the concept of semiotics from different scholars and converge the common parts of different views. Cooperate with modeling elements such as model, color, materials and texture, and convert the goods from the intangible concept to the tangible entity. As to the format and levels, they are influenced by the cultural concept which came from the theory of three layers of culture. Finally, integrate the three concepts and construct knowledge base with easy operation and convenient application.

Step III: Collected the local materials and natural, production, human, cultural and landscape resources, and converge the local literature. Input the local resources and materials to the cultural and historical design symbol knowledge base.

Step IV: Take a local resource as an example and input the information regarding the landscape "Jiujiu Peak", to construct a complete knowledge base.

Step V: Complete the construction of cultural and historical design symbol knowledge base; follow the procedure of developing cultural and creative products and complete the development model (Fig. 4.5).

4.3.1 Step 1: The Original Concept of the Knowledge Base of Cultural and Historical Design Symbols

1. The original concept of transforming local resources into knowledge bases

Arthur Andersen, an industry management consultancy, divides knowledge into four concepts known as "knowledge assets (Andersen, 2000)." This concept will be transformed into the original concept of data input of Caotun's natural, production, human, cultural and landscape resources. The following divides the concept of how cultural and historical materials become local intellectual property into four processes:

(1) Cultural and historical data: namely primary or original data (Wu et al., 2010), through the collection of Caotun's culture, history, cultural relics and other documents.
(2) Design information: take the obtained information as the subject, and organize it with a purpose to convey a certain message (Andersen, 2000; You, 2001). Through the information about Caotun's culture, history, cultural relics, etc., the design information that can be conveyed is sorted out.
(3) Creative knowledge: By analyzing the information about the Natural, Production, Human, Cultural and Landscape, we can master the ability to take advantage of the opportunity. Therefore, the design information is summarized and analyzed to identify elements such as shape, color, texture, material, function, and spiritual connotations.
(4) Intellectual property: This is based on knowledge (Andersen, 2000), using symbolic elements such as shapes, colors, textures, materials, functions, and spiritual connotations. Use these elements, combined with design experience

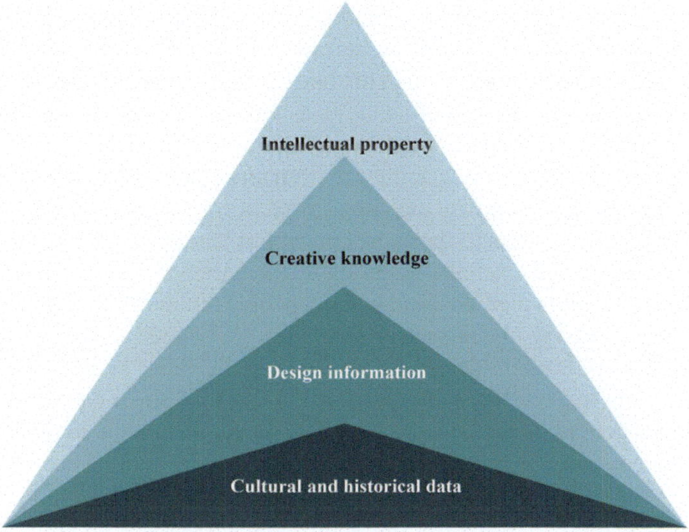

Fig. 4.6 Transformation of local cultural and historical data into local intellectual assets

and abilities, to fulfill the goal of creating valuable local cultural creative products. The purpose of the above four processes is the establishment of the conceptual form of Caotun's local cultural and historical design knowledge base. As shown below in Fig. 4.6 "Transformation of Local Cultural and Historical Data into Local Intellectual Assets."

4.3.2 Step II: Build Knowledge Base Horizontal and Vertical Axis Design and Format Content

1. The modeling elements of the knowledge base

The cultural and historical design symbol knowledge base in this study can convert the modeling elements to tangible forms such as model, color, material, texture and function, etc. The following is a discussion of relevant academic literature on how elements of design are constructed in cultural creative products. The concept of design originates from the Bauhaus School of Design established in Germany in 1919, so the word "design" has no proper original text in English, and Gestalt in German means "unified whole," that is, the Gestalt explored in Gestalt psychology (Lv, 1984). The three-dimensional shape Form is caused by the movement of the surface, and the flat design of the Shape is caused by the movement of the line (Hiroshi, 1967). However, designing actually covers two levels of meaning: to create and to shape. The former is a verb, and the latter is also a noun. Through human will and cognition or natural laws, the process of completing a shape through some media is called "design" (to design).

Design can be divided into both broad and narrow senses. In the broader sense, the interpretation is about transforming the information or situation received from the outside into meaningful shapes through the human visual organs; while the narrower sense of design refers to the idea of using ideas within the human heart or external, natural laws to create a shape through some medium. Therefore, designing is the matter of creating a "shape." In fact, the act of creation should include the meanings "work" and "production"; and the meaning of "creation" in the term design (the *zao* of *zaoxing* in Chinese) refers to the same concept of "creation" in the term creation (*chuangzao*). Design is the act of expressing the meaning of the existence of objects by using various forming methods, and making people remember various "existences in the form of expressing meaning (Lin, 2001, 2014)." The Ministry of Education's Mandarin dictionary explains designing "integrating the various parts into a structure or form according to experience and purpose." Therefore, the meaning of the word design includes not only visual qualities such as appearance, light and shadow, color and texture, but also involves an in-depth exploration of the item's functions and how to produce it (Sunkiss, 2003). The term "design" can be explained using the following points: The first level of design is the action and process of creating the shape. The second level of design is the specific shape composed of various mechanisms and functions. The third level of design is composed of shape, color, material and texture, emphasizing the overall image after integration (Yang, 1997). Based on the above theories, the design elements of the knowledge base must be composed of shape, color, material, texture, etc. through the shape elements, and integrated into the function and user interface of the product. The formation of local cultural and creative products is no exception and they can form the tangible goods through model, color, material, texture and function.

2. The interpretation forms of the knowledge base

This interpretation on the semiotics universality in the literature aims to summarize the denotative meaning and connotative meaning of symbols, which is conducive to analyzing the local characteristic resources, deconstruct or construct local cultural elements and develop the special goods. To sum up the theories of interpreting the symbols by the scholars, we found the following results. Pierce explains the symbols from three aspects, namely, icon, index and symbol. According to the Swiss linguists Saussure, the symbol has paradigmatic axis and syntagmatic axis, similar to the vertical and horizontal axis. And he proposed signifier and signified to explain the denotative and connotative meaning of symbols. The American philosopher Morris put forward three fields of syntactics, semantics and pragmatics, to interpret the denotative and connotative meaning of symbols. Roland Barthes, the French scholars, refers to the first level meaning of a symbol as denotation; the second level of meaning: hidden meanings, myth, symbolic, in which he use level to interpret symbols. The Italian scholar Umberto Eco interpreted symbols to language rules theory, mark production theory, connotative meaning and denotative meaning, which mean the same as those we put forward before. Finally, scholar Katsumi Hoshino compared the symbol to the commodity value structure and explained the commodity value from its denotation and connotation. The above interpretation of the symbols from different views of semiotics is helpful for the construction of the knowledge base (Table 4.6).

By a comprehensive view of common points of different theories of semiology, it is found that most were signs' external connotation, such as modeling, color, texture, material and functions of cultural and creative products, and the signs' internal connotation, explanation of the symbolic cultural connotation of the story and history of the cultural and creative products.

3. Stratification of the knowledge base

Culture is an activity of differentiating and understanding the life of different human groups, which includes three levels: metaphysical level where there are thinking activities and language; intermediate level where there are institutions about co-existence and interaction among people, and the physical level where there are tools used by people and concrete and visible things (Yang, 1998).

Table 4.6 Theories proposed by semiotic researchers (Wang, 2020)

Duration	Scholar	Theory	Methods for interpreting symbols
1839–1914 Father of pragmatism and symbolism	Charies Peirce	Three elements of symbolism: symbol, interpretation, and object	Three types of symbols: index, symbol, and icon
1857–1913 Swiss linguist	Ferdinand de Saussure	Trunk axis, neighboring axis, symbolic tool, symbolic meaning	Interpreting vertical and horizontal thinking as well as denotative and connotative meanings of symbols
1901–1979 American philosopher	Charies Morris	Three properties of symbols: Syntactics, Semantics, and Pragmatics	People, Objects and Elements, interpretative symbols
1915–1980 French scholar	Roland Barthes	First-layered meaning: explicit meaning. Second-layered meaning: implication, myth, and symbol	Two-layered interpretative symbols
1932 Italian scholar	Umberto Eco	Linguistic principles, theory about symbol-creating	Interpretative symbols on denotative and connotative layers
1988 Professor of management information department and author of marketing and semiotics	Katsumi Hoshino	Structure of commodity values	Use and symbolic value of products

(1) The first layer "a culture's metaphysical": refers to spiritual culture, including art, religion, and ethics, including decoration, painting, sculpture, ballads, stories, and dance, as well as attitudes and behaviors toward supernatural states and states of divine existence; interpreted as "Intangible Conceptual Interpretation (Yang, 1998).
(2) The second layer of "a culture's intermediate": refers to community culture, examples and lifestyles, social relations, customs, economic status, politics (power distribution and operation), formal laws, rules of design, interpersonal relationships, ultimate beliefs, myths and legends, etc.; interpreted as "The Life of a Realistic Character Experience (Yang, 1998)."
(3) The third layer "a culture's physical": refers to material culture, such as food, clothing, housing, utensils, artistic relics (tangible and intangible), means of production, fragments and other things that are clearly related; interpreted as "A Tangible State of Symbolic Transformation." (Fig. 4.7) (Yang, 1998).

Through discussion of the "three-leveled culture", we can divide the Caotun cultural and historical design symbol knowledge base into two levels, namely "cultural concreteness" and "cultural connotation", to construct the knowledge base's form and content, which include:

(1) "The culture's metaphysical connotation" in this study is the same with the concept of cultural concreteness, which includes tangible forms, such as color, texture, modeling, surface ornamentation, lines, treatment of particulars, functions, operability, etc.

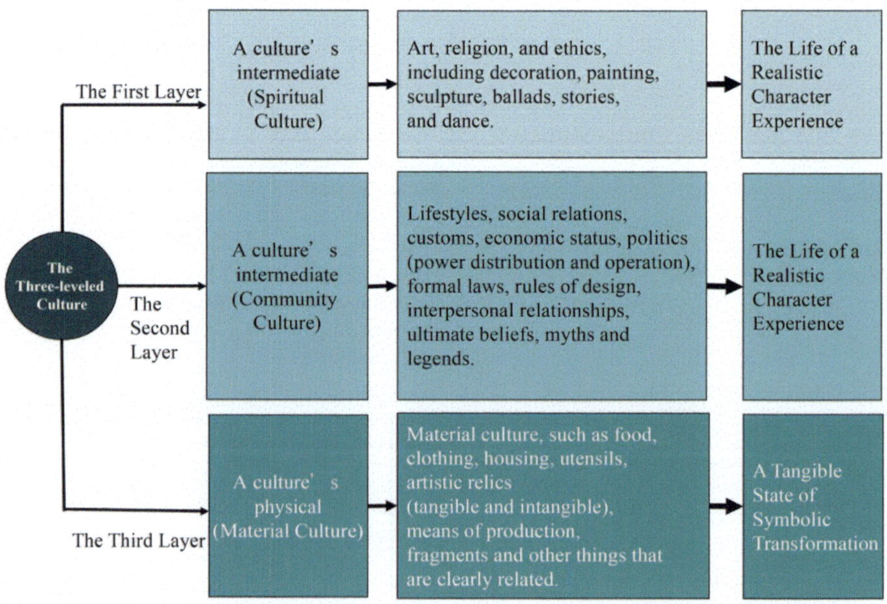

Fig. 4.7 The three-leveled culture (Yang, 1998)

(2) "The combination of the culture's intermediate level" and "physical level" is the same with the concept of cultural connotation, which refers to the invisible cultural attributes, such as the spirit, story, history and uniqueness, etc. that are bestowed in the local cultural and creative products.

The vertical and lateral thinking model partitioned within the knowledge base is based upon the contents of the three components of: knowledge base design elements, symbolic interpretation format, and the three layers of cultural analysis. Through this model, a simple and easy to use method of searching for content within the knowledge base is established (Fig. 4.8).

As for construction of horizontal and vertical content of the knowledge base, there are five levels in the horizontal axis, the first level is appearance, cultural concreteness and connotation; the second level is culture, content and forms; the third level is turning characters into graphs; the fourth level is coming up with more design ideas after vertical and horizontal thinking about expressing culture in characters and graphs; the fifth level is materializing the concept through the from-point-line-surface-to-body process. The vertical axis includes two levels: the first level is "culture's denotative meaning", whose content comprises pictures, material and its use, choice of color, external form of the picture, function and interface design, style, design vocabulary, idioms, three-dimensional composition and design structure, etc. The second level is "culture's connotative meaning", which contains cultural and spiritual connotation and cultural folklores and stories (Fig. 4.9).

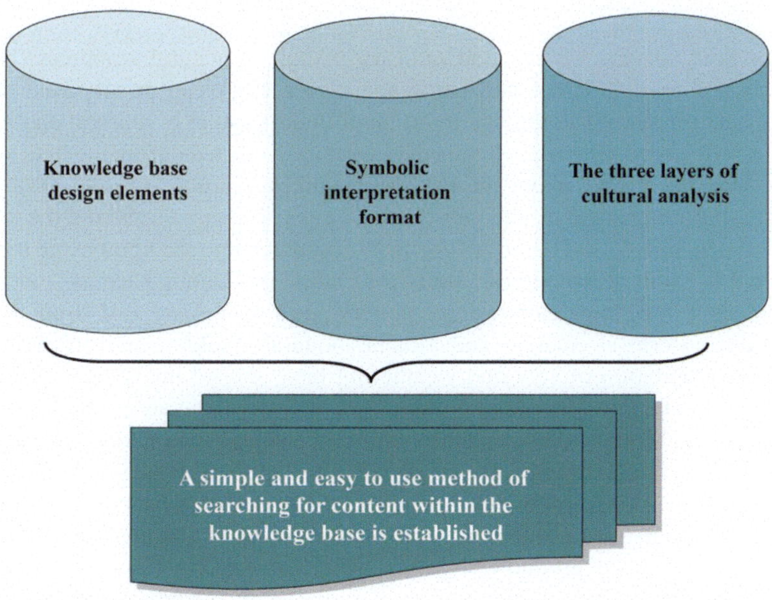

Fig. 4.8 Integrate the three groups of concepts to build a knowledge base in the form of lateral and vertical thinking

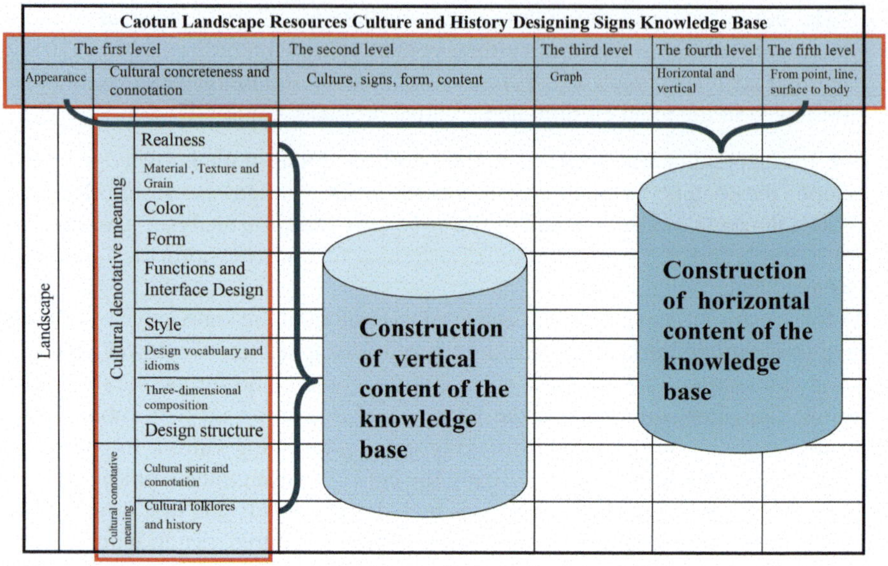

Fig. 4.9 Divide the content of the knowledge base into sections focused on either vertical or horizontal thinking

4.3.3 *Step III: Select Local Resources*

Through field survey, we selected industry resources, natural resources, human resources, cultural resources and landscape resources (Weng & Miyazaki, 1996) that can best represent Caotun and input them into literature of cultural and historical knowledge base. The following is the literature collected during the field survey and the integrated table about the natural resources, landscape resources, human resources, cultural resources and industry resources. The aim of making the table is to input types, photos and details of the local resources into the knowledge base.

Table 4.7 "Natural Resources," study the "sites" of Caotun, such as stone axes, hoes, polished stone tools, stone knives, pottery pieces, and so on; and "Jiujiu Peaks" formed due to crustal changes and violent mountain-building movements; rapid erosion, accumulation of soot and weathering have shaped the special geological topography of Huoyan Mountain; looking at the Shuangdong area from Tuchengli, the "Tombstone Mountain" looks like a tombstone, and the mountain is shaped like a phoenix crown; the "Dahu Mountain," which is only 385 m above sea level halfway up the mountain. This low mountain next to the metropolis has become a hiking trail for locals due to its gentle slope and nice views; "Jialao Mountain," east of Beitou-She, protrudes from flat ground with peaks. The "Time-honored Camphor Tree" tree on the Zhongtan Highway going to Nanpuli and entering the hillside by Pingding is as much as 1,600 years old, 32 m tall, has a circumference of 8 m, and a canopy which stretches across 800 square meters. The tree is both tall and magnificent. "Production resources" includes the products of Caotun symbolizing "straw sandals," and others include "rice," "betel nuts," "grapes," "lychees," "white pomelos" and "barley seeds." "Human resources" originated from the Han ancestors who immigrated to Caotun Town in the Qing Dynasty, and the "four surnames Hong, Li, Lin, and Jian" were the mainstream. There are many famous doctors under the eaves of Furong Hall, and they are called "the family of doctors in Caotun." Huang Chunfan in Caotun pioneered the construction of Longquan Canal and contributed to the development of agriculture, and is known as "the father of Caotun Water Canal." Caotun's "Cultural resources" is related to the story of the place names in Caotun, such as: "Lugang Danpushe," "Lin Shuangwen led his troops to abandon straw sandals to form a small mound," "Zheng Chenggong led his troops to pass through these abandoned shoes and turned into a mound," "The land around Caotun is shaped like a straw sandal," and so on. Representatives of Caotun's "Landscape resources" include the "Eight-diagrams Kiln," "Shuangshi Suspension Bridge," "Furong Hall," "Dengying Academy," "Old Construction on the Caotun Old Street" and the "National Taiwan Craft Research and Development Institute." The above five resources of Caotun are all collected from the field investigation and literature collection in this book. The purpose is to select the resources which best represent Caotun from among the collection of various resources.

There are eight rows marked with a "◎" in Table 4.8, indicating that this resource will be selected as an example for the construction of a knowledge base of cultural and historical design symbols after field investigation and document collection. For natural resources section the representative natural wonder "Jiujiu Peaks" and the majestic "Tombstone Mountain." For production resources section, the "straw sandals," which are representative of the locality, are chosen. To this day, local promotions still use straw sandals to represent Caotun. Human resources include "Caotun Scholars." Since olden times, the construction of local academies in Caotun has had a positive influence on the cultural and educational atmosphere, and cultivated many scholars and academic-focused families. For cultural resources section, choosing the story of "Lugang Danpushe," where those walking from Lugang to Puli would pass through Caotun and replace their shoes, forming a mound, is important to the history of the township and thus was chosen. For landscape resources, the special scenery of the "Eight-diagrams Kiln" in Caotun, the "Furong Hall" of the family of doctors in Caotun, and the third-level historic site, an ancient academy in Caotun, the "Dengying Academy."

1. Local resource input knowledge base

The eight representative resources of Caotun under the division of natural, production, human, cultural and landscape all have background resources for the pre-production process of local resources being converted into data input for the knowledge base. Figure 4.10 begins the "knowledge input" process, which summarizes local resources through field investigations, historical documents, interviews with experts, scholars, and writings. However, the "knowledge establishment" is divided into a format structure through the concept of a cultural and historical design symbol

Table 4.7 Summary table of natural, production, human, cultural and landscape in Caotun

Resources of Caotun		
Type of resource	Picture of resource	Details of resource
Natural resources		Stone axes, axes and hoes, polished stone tools, stone knives, and pottery sherds were found at the site (mainly red and gray earthenware, with cord and checkered patterns on the surface)
		Jiujiu Peak: Also known as the Flaming Mountains, it is a scenic spot in local Caotun
		Tombstone Mountain: Tombstone mountain is shaped like a tombstone, and the mountain is shaped like a phoenix crown
		Dahu Mountain: Dahu Mountain overlooks Bagua Mountain and Caotun Street. On the mountain there are: Taiwan Lei Zang Temple, the temple of the Chinese True Buddha Esoteric Association, the Nineteen Dragon Pillars of Taoism, Ci De Temple and Ci Hui Hall, and Guanghui Culture and Education, the headquarters of Yiguandao believers, as well as the Guanghui Cultural and Educational Memorial Hall, along with two land temples

(continued)

Table 4.7 (continued)

Resources of Caotun		
Type of resource	Picture of resource	Details of resource
		Jialao Mountain: The term "kalou" is a term used by the local Plains Indigenous tribe residing in the area. The pronunciation is "kalou," which means bamboo. Han peoples first translated it into Hokkien as "kiáu-ló," and later changed it to ka-ló." According to the Changhua County Annals of the 1826 edition, "Xuanlong Mountain, east of the Beitou settlement, protrudes from the flat ground and has peaks. The local language refers to it as: kiáu-ló, bamboo ware. The top of the mountain is named after the pictograph." Looking east from the bank of Bâu-lô (Maoluo) River, the dome peak seen in the middle of Oo-khe (Wuhsi) is today's Jidai Mountain
		Time-honored Camphor Tree: Located about six kilometers southeast of Caotun, there is a mountain camphor tree, with dense branches and leaves towering to the sky. The local people respectfully refer to it as "Zhangshugong," and most people also call it "Qigu Sacred Tree." According to the informative tablet erected by the Nantou Forest District Office and the Nantou County Ecological Conservation Association in the 1990s: the "Qigu Sacred Tree" is as old as 1,600 years old, 32 m tall, has a circumference of 8 m, and an 800 square meter canopy. The tree is tall and magnificent
Human resources	洪李 林簡	Most of the Han immigrants to Caotun were from Zhangzhou, and the four surnames "Âng (Hong), Lí (Li), Lîm (Lin), Kán (Jian)" are the most dominant
		Caotun scholars: Chuang Wen-wei, Hong Chung-ying, Hong Yueh-chiao, Hong Lian-kui and Li Chang-chi
		Traditional family of doctors in Caotun: Hong Hsun-yao, Hong Jui-sung, Hong Hsun-hsin and others are all famous doctors under the eaves of the Hong family in Furong Hall, and they are known as the family of doctors in Caotun

(continued)

Table 4.7 (continued)

Resources of Caotun		
Type of resource	Picture of resource	Details of resource
		Father of Caotun Water Conservancy—N̂g Tshun-huân (Huang, Chun-fan): In the 10th year of the Taisho Emperor's reign (1922), the construction of the Longquan Canal was the first major project to promote the development of agriculture in the Caotun region
	 	During the Qianlong and Jiaqing years, predecessors went to Lugang, Changhua, Kuaiguan, Shekou, Xianzhuang, Beitou, and Tsháu-ê-tun, and to the east to Puli-she, where the settlers, porters, and business travelers were in Caotun at the beginning. The place where old and new straw sandals were replaced, and the real incident of straw sandals accumulating into small mounds over time. "Lugang Danpushe" has become a popular story retold by residents during leisure time
		Zheng Chenggong led 3,000 troops to fight against the Douwei Longan (Pazeh) Indigenous tribe. After this, they abandoned their shoes, creating a mound of straw sandals in the process. In addition, it is also said that Zheng Chenggong's troops were going to enter Puli to attack indigenous tribes known as the Hong. Originally, they had to pass through Guoxing, but because of cholera outbreaks in the local area, they altered their route, passing through Cao Xie Dun instead. After passing, the troops rested here and changed their straw sandals. The straw sandals that were taken off were piled up into hills, so the area was called "Cao Xie Dun" (Straw Sandal Mound)
		In the fifty-third year of Emperor Qianlong's reign in the Qing Dynasty, when Lin Shuangwen was fighting against the Qing Dynasty, the Qing soldiers who were conquering the western part of the island passed by here and abandoned straw sandals in piles. At the end of the Ming Dynasty, Zheng Jing led his troops to attack the "Douwei Long" indigenous tribes (part of an Atayalic ethnic group) and three thousand officers and soldiers passed by, and also abandoned their straw sandals in piles. Thus, forming the origin of the place name "mounds formed by abandoned straw shoes (sandals)."

(continued)

Table 4.7 (continued)

Resources of Caotun		
Type of resource	Picture of resource	Details of resource
		During the Qianlong period, "In the Shadow of Cao Xie Dun (Straw Sandal Mound/Mountain)" appeared, and the topographic map of the area was shaped like a straw sandal, hence its name
Production resources		In the past, the town was dominated by agriculture, of which rice was the most productive in the central area. At that time, the settlers, porters and business travelers who went to Lukang and Puli liked the exquisite and practical straw sandals woven by artisans in Caotun
		Caotun is rich in agricultural products, such as rice, Shuangdong betel nut, Pinglin grape, Pinglin lychee, white pomelo, coix seed and pear
Landscape resources		Dengying Academy: In the Qing Dynasty, martial arts masters appeared here, and the site was listed as a national third-level historic site. This ancient site is well preserved, and there are clipping totems such as Qilin and dragons on the roof and eaves
		Furong Hall: Since Mr. Hung Te-chung likes to plant flowers and trees, and he also expects the prosperity of the Hung family, he quotes Wang Xizhi: "Cultivating mulberry fruit, now it will be prosperous"; Furong Hall's house name is self-proclaimed

(continued)

Table 4.7 (continued)

Resources of Caotun		
Type of resource	Picture of resource	Details of resource
		Old Construction on the Caotun Old Street: Residents say, "It Hok, Nng Lu, Sam Bangkah, Su Bakthau" Beitou is now the old street in Caotun's urban area. There are exquisite cultural symbols from China, Japan, and Europe adorning the many historic houses along the old street
		Eight-diagrams Kiln: The shape of the kiln body is oval, and the roof is octagonal, similar to the octagonal-shaped eight trigram symbols representing Yin and Yang concepts popular in local culture. The kiln continues to burn day and night, just like the cyclical, never-ending nature of the eight trigrams (Bagua). Thus, the kiln is called the Bagua Kiln (Eight Trigram Kiln)
		Shuangshi Suspension Bridge: The Shuangshi Suspension Bridge spans the Wuxi River, connecting Pinglin-li and Shuangdong-li. With a total length of 360 m, it is the most complete and longest suspension bridge in the province. The flaming red colored suspension bridge is very exciting to the eye. The bridge allows visitors to take a stroll between the greenery of the mountains and the shimmering, clear river. Looking down, one may observe the gurgling water, which is so clear that one can see the bottom
		National Taiwan Craft Research Institute: The craft center is located at the key point between Sun Moon Lake, Cingjing Farm, and Formosan Aboriginal Cultural Center. Children play on its vast green lawns, collecting acacia beans under the thriving old trees, making eternal memories. The craft exhibition hall is majestic. Modern architecture has become an important landmark in central Taiwan. It displays the latest creations and products of domestic craftsmen and artists, which is worth carefully appreciating

Table 4.8 Selection of eight representative resources in Caotun

Eight representative kinds of resources (natural resources, production resources, human resources, cultural resources and landscape resources, etc.) obtained through field survey				
Natural resources	Production resources	Human resources	Cultural resources	Landscape resources
1. Historical site	◎1. Straw sandals	1. Four Large Family Names: Hung, Li, Lin and Chien	◎1. Lugang Danpushe	◎1. Eight-diagrams Kiln
◎2. Jiujiu Peak	2. Rice	◎2. Caotun scholars	2. Folklore about Caotun being named so for Lin Shuangwen discarding his old sandals here	2. Shuangshi Suspension Bridge
◎3. Tombstone Mountain	3. Betel nut	3. Traditional family of doctors in Caotun	3. Folklore about Caotun being named so for Cheng Chenggong discarding his old shoes here	◎3. Furong hall
4. Dahu Mountain	4. Grapes	4. The Caotun pioneer ditch-digger	4. Sandle-like geography of Caotun	◎4. Dengying academy
5. Jialao Mountain	5. Lychee			
6.Time-honored Camphor Tree	6. White Shaddock			5. Old construction on the Caotun Old Street
	7. Coix seed			6. National Taiwan Craft Research and Development Institute

knowledge base, and textualizing pictures, and pictorializing text. After the knowledge base is established, through "knowledge extraction," the knowledge base is built into the process of cultural creative product design and production, extracting data from the knowledge base to use during product design. Finally, "knowledge application" is the conceptual form of the future cultural and historical design symbol knowledge base, which benefits from designers or local use.

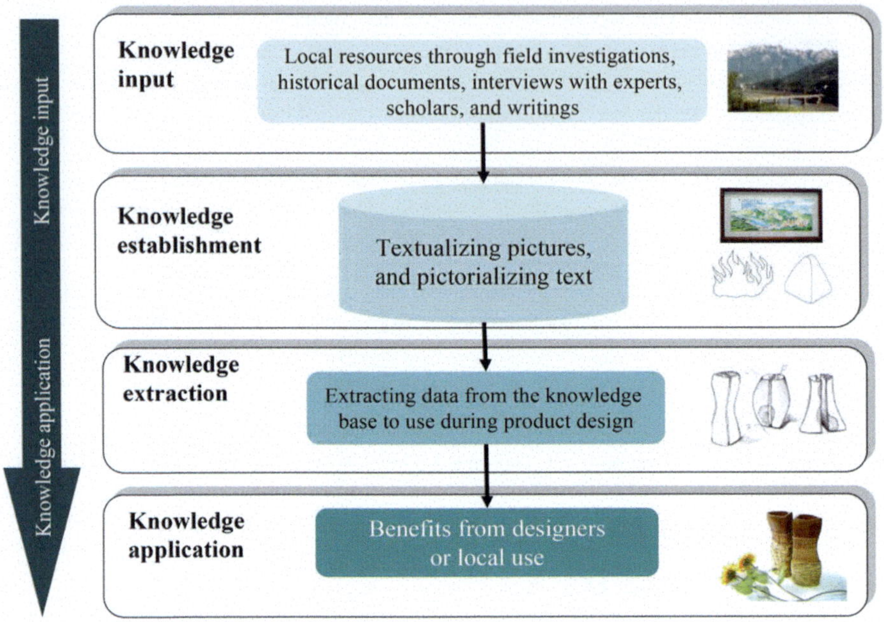

Fig. 4.10 Input and application of local knowledge base resources

4.3.4 Step IV: Complete the Construction of the Knowledge Base Form

Because it is impossible in this study to explain every knowledge base about natural, production, human, cultural and landscape resources in Caotun, the Jiujiu Peak is selected here as an example to show how to construct knowledge base for cultural and historical signs designing. Jiujiu Peak, representative of local landscape of Caotun, is the national natural ecology preserve area and one of the wonderful sights in Taiwan. In the following Table 4.9, the cultural and historical design symbols knowledge base about Jiujiu Peak is divided into five levels, with detailed explanation as follows:

1. The first level is about "realness", "cultural concreteness" and "cultural connotation". Realness refers to the landscape resources, like Jiujiu Peak. "cultural concreteness" and "cultural connotation" include "cultural denotative meaning", which contains appearance, material, texture and grain, color, form, functions and interface design, style, design vocabulary, idioms, three-dimensional composition and design structure, etc. and "cultural connotative meaning", which is the cultural and spiritual connotation, cultural folklores and stories.
2. The second level is where the "cultural denotative meaning" is further divided into appearance, material, texture and grain, color, form, functions and interface design, style, design vocabulary, idioms, three-dimensional composition and design structure, etc.; and the "cultural connotative meaning", including the

Table 4.9 Jiujiu Peak's cultural and historical design symbols knowledge base

Caotun Landscape Resources Cultural and Historical Design Symbols Knowledge Base							
The first level				The second level	The third level	The fourth level	The fifth level
Realness		Cultural concreteness and connotation		Culture, signs, form, content	Graph	Horizontal and vertical	From point, line, surface to body
Landscape	Jiujiu Peak	Cultural denotative meaning	Appearance			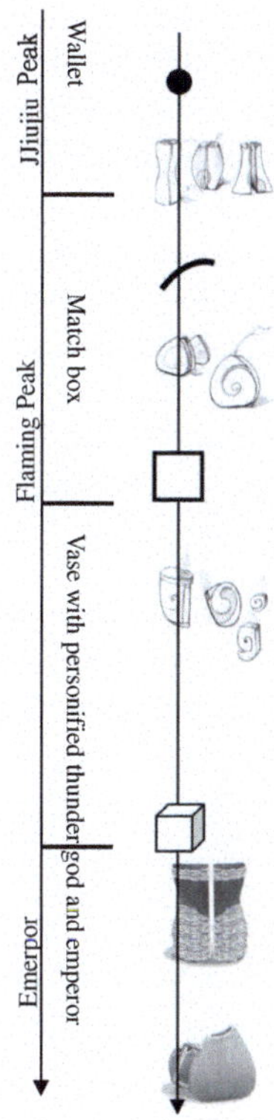	
			Material , Texture and Grain	Vegetation, forest			
			Color	Before 921: bright green; after 921: yellow Color of natural sights such as bleak mount			
			Form	Taking the form of a flame, and layers and layers of steep peaks			
			Functions and Interface Design	Folklore says that Lin Shuangwen lead rebellion in Qing dynasty; he buried military silver here for later rebellion. On his death he said people being blessed will find the treasure, but till now no body has found the treasure yet.	18 boxes of military silver 		
			Style	Peak-style, flame-style			
			Design vocabulary and idioms	Jiujiu Peak			
			Three-dimensional composition	Inorganic angle, triangle formation, structure of overlapping triangles			
			Design structure	Different view from different perspectives			
		Cultural connotative meaning	Cultural spirit and connotation	Emperor Peak			
			Cultural folklores and history	According to Qing dynasty folklore, the Mountain originally had 100 peaks and wanted to be crowned the emperor, but the god disagree and used thunder to break one peak, so only 99 peaks are left.	100 - 1 = 99 Thunder God striking the emperor		

cultural and spiritual connotation, cultural folklores and stories, are described in detail in writing.

3. The third level is picturization, which means using graphic or digital form to describe the "cultural denotative meaning", namely, appearance, material, texture and grain, color, form, functions and interface design, style, design vocabulary, idioms, three-dimensional composition and design structure, etc. and the "cultural connotative meaning", including the cultural and spiritual connotation, cultural folklores and stories.
4. The fourth level is the "vertical and horizontal" concept, which is the extension from that of the semiologist Saussure, where relationship between paradigmatic axis and syntagmatic axis is discussed, and this is the vertical and horizontal thinking method frequently used in designing. In the horizontal thinking direction, Jiujiu Peak (in vertical direction, we recalled that treasure is buried under the mountain in Qing dynasty, so the mountain reminds us of a wallet) is also called Flaming Mount (by vertical thinking the flaming Mount, covered by fire, makes us think of a match box); it still retains the title of Emperor Peak (The folklore of personalized Thunder God striking the Emperor makes us think of an ornamental vase in the vertical thinking.).
5. The fifth level is "from-point-line-surface-to-body process", during which the culture-based creative products" draft drawing is produced step by step along the above four levels: description in writing or graphic form, vertical and horizontal thinking, etc.

Figure 4.11 shows the achievements after completion of form of the cultural and historical design symbols knowledge base about natural resources, human resources, production resources, cultural resources and landscape resources, wherein all the above resources can increase files in the knowledge base successively for the local designers to develop.

4.3.5 Step V: Application and Procedure of Knowledge Base

After constructing the cultural and historical design symbol knowledge base, the first thing we faced was to consider how to use the content of the knowledge base, and whether the concept of the development mode in this study can be put into practical use. The following is a detailed description of the methods to apply the knowledge base and the procedures concerning cultural and creative products modeling, which refers to the procedures of physical product development in industrial design, leading itself in the concept development process (Liu, 2007). The reason for that is because the industrial design process has such advantages as rigorous inspection at each stage and quantization. Besides, it can infuse cultural and creative products of local industry into the concept of commercialization development, and help achieve the goal of "Culture Commodity and Commodity culture". That's why the development concept of the industrial design is introduced into the procedures of cultural and creative

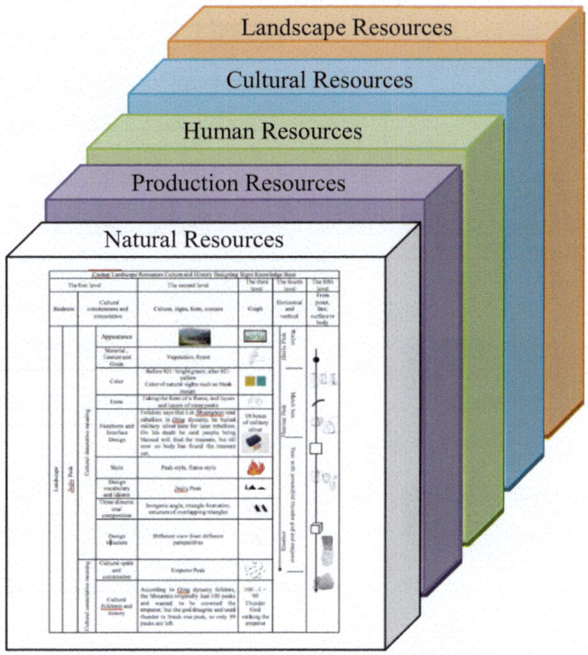

Fig. 4.11 Cultural and historical design symbols knowledge base about natural, production, human, cultural and landscape resources

products modeling in this study. The following are the four modeling procedures for using the knowledge (Fig. 4.12).

1. The 1st step—defining the design guidelines for the knowledge base: to select the landscape resource of Jiujiu Peak for the cultural and historical design symbol knowledge base and extract the modeling, color, texture, material, connotative meaning, spirits of legends and tales to further determine the design conditions for cultural and creative products.
2. The 2nd step—developing ideas: after determining cultural and creative products conditions, draw the "Idea Sketches" of the goods; and choose one of them to make a three-view drawing of the cultural and creative products: front view, top view, side view and space diagram.
3. The 3rd step—screening and final election: making proportional sketches with materials in accordance with personal preference, inspecting whether the proportions of the three-view drawing comply with each other; taking the feasibility of the products into consideration such as: basic human factors, physics, mechanics etc.
4. The 4th step—creating final goods: drawing an accurate full-size draft, creating the next "final goods manufacture" procedure in accordance with the full-size draft, and naming them after the completion of the final goods (Table 4.10).

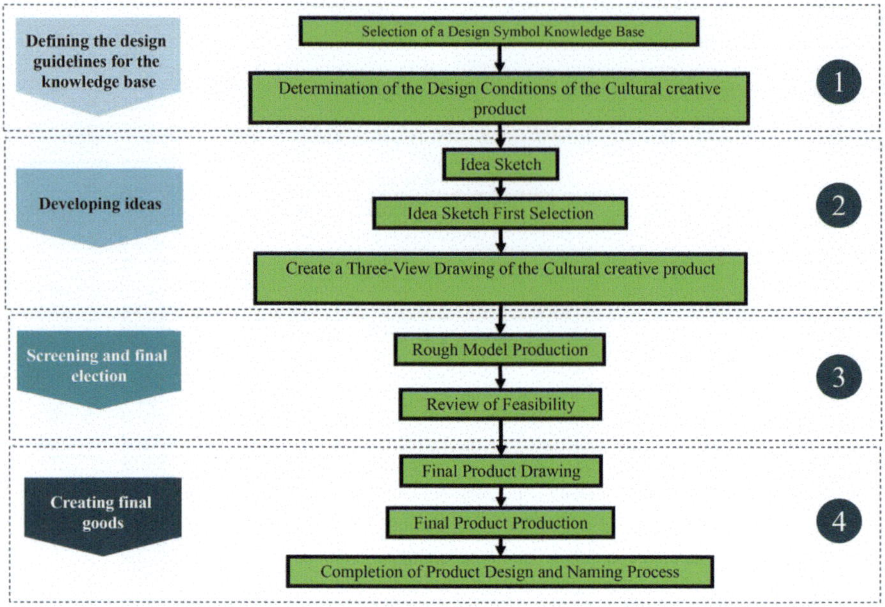

Fig. 4.12 The application and process of the knowledge base

After a detailed description of the cultural and creative products creating process, it is still necessary to inspect whether the "cultural concreteness" and "cultural connotation" of the knowledge base, comply with the spirit of local elements. The following Table 4.11 is an original work to verify the cultural denotative and connotative meaning of the knowledge base.

4.4 Results of Cultural and Creative Product Design as Viewed in Knowledge Base

The following two cultural and creative products: the Jiujiu Peak and the Thunder God Striking the Emperor, were created in accordance with the cultural and historical design symbols knowledge base. The first product Jiujiu Peak is developed in the direction of the denotative meaning, with a shape of "as a ridge from the side, cross-peak"; the color derives from the earth's yellow color after the 921 Earthquake; texture indicates the momentum of the loess; function is related to the fact that the Jiujiu Peak was listed as a natural conservation zone, so it is created as a vase. The second product Thunder God Striking the Emperor is about connotative meaning, representing the tales on the Jiujiu Peak. According to the folk tale in the Qing Dynasty, there were originally one hundred peaks in the mountain where the Emperor was born. However, the God was not satisfied with it, so he ordered the Thunder God

Table 4.10 Design and manufacturing processes of creative products

Development steps	Design flow chart	Instructions for each step
Selection of a design symbol knowledge base	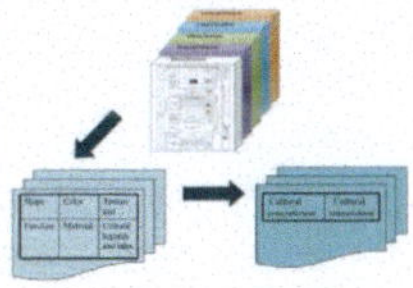	First, select an example from the "Cultural and Historical Design Symbol Knowledge Base," such as the landscape resources of JiujiuPeaks, extract the cultural denotation (signifier), and the connotation (signified)
Determination of the design conditions of the cultural creative product	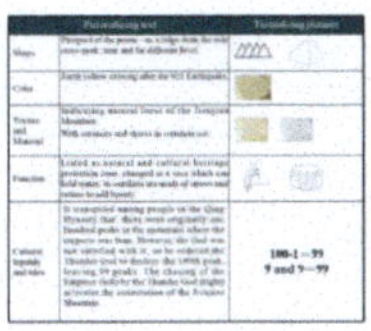	Extract the cultural denotation and connotation from the knowledge base, and decide the design direction and category of the product
Idea sketch		1. Begin graphical and shape design 2. The more brainstorming the better
Idea sketch first selection		1. Express the design more clearly. The rougher the sketches, the better 2. Select three draft sketches for further development
Create a three-view drawing of the cultural creative product		1. Match molding components 2. Review production and processing technology 3. Choose one three-view drawing draft for further development of rough and fine models
Rough model production		A proportional rough model does not have a restriction on the materials used in its construction, for example: paper clay, PU foam, pottery, clay, modeling clay, paper boards, anything which can convey the overall shape and design can be used

(continued)

Table 4.10 (continued)

Development steps	Design flow chart	Instructions for each step
Review of feasibility		1. Basic considerations of human factors, physics, and mechanics 2. Methods of producing and processing, materials of the model
Final product drawing	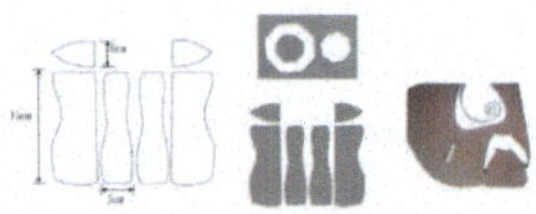	Draw a 1:1 stereogram
Final product production	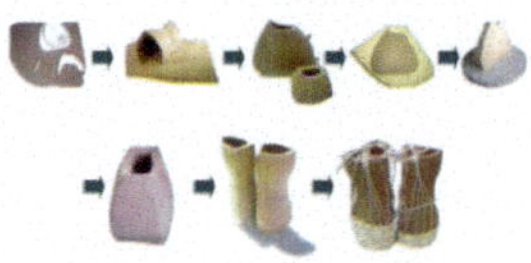	1. Consider the color and processing characteristics of the surface treatment 2. Consider the assembly of heterogeneous materials
Completion of product design and naming process	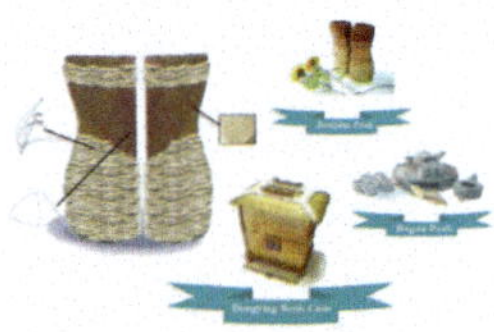	Giving the final cultural creative product a name which reflects or is related to its cultural spirit

to destroy one of the 100 peaks, with 99 peaks left. The chasing of the Emperor (left) by the Thunder God (right) activates the connotation of the Jiujiu Peak, because the folk tale draws the distance between the products and the public closer. To assist creation with the knowledge base can realize the connotative and denotative meaning of the local goods. For example, with the help of the knowledge base, we can reduce unnecessary, trivial and complicated information, and directly focus on the modeling, color, texture, material, function of the local culture and create in a scope close to the connotative meaning to shorten the development time on the one hand, and master the symbols and elements of local culture on the other hand (Table 4.11).

4.5 Results of Knowledge Base Feasibility Research and Analysis

Whether or not using semiotics to construct a knowledge base, and integrating it with local industrial resources, will achieve the goal of enabling foreign and local product development designers to use the knowledge base as a reference for information, prompted this study to condense the characteristics of local elements when designing cultural creative products. However, the feasibility of a knowledge base

Table 4.11 Review on design results of creative products

Product name	Comparison between "cultural concreteness" and "cultural connotation"
The first product: Jiujiu Peak	Cultural concreteness (cultural denotative meaning)
	Shape: prospect of the poem—as a ridge from the side, cross-peak, near and far different level
	Color: earth yellow existing after the 921 Earthquake
	Texture: indicating natural loess of the Jiujiu Peak
	Material: with ceramics and straws in common use
	Function: listed as natural and cultural heritage protection zone, changed as a vase which can hold water; its outskirts are made of straws and rattans to add beauty
The second product: the Thunder God Striking the Emperor	Cultural connotation (cultural connotative meaning)
	Cultural legends and tales: it transpired among people in the Qing Dynasty that: there were originally one hundred peaks in the mountain where the emperor was born. However, the God was not satisfied with it, so he ordered the Thunder God to destroy the 100th peak, leaving 99 peaks. The chasing of the Emperor (left) by the Thunder God (right) activates the connotation of the Jiujiu Peak

usage by "foreign and local product designers" remains unknown. The local product development is dominated by members of the Cao Xie Dun Local Cultural and Educational Association (hereinafter referred to as the Cao Xie Dun Association), because most of the local cultural creative products in Caotun are in charge of the design department of the association. The following will count the students from the Industrial Design Department of Tunghai University, the non-local designers, and the local designers, members of the Cao Xie Dun Local Culture Association, and create a questionnaire to analyze the feasibility of using the knowledge base to guide designers in their efforts to develop products.

4.5.1 Definition of Terms Related to Statistics

In this study, detailed data were obtained by statistical analysis through SPSS software after the questionnaire survey, and displayed and analyzed using pie chart percentage, cross analysis, factor analysis, reliability and validity analysis, descriptive statistics, along with other analytical methods. The following is the definition of statistics related terms and the application of analysis methods.

1. The semiotic theory of Charles Morris

Using Bivariate Data, it is often possible to investigate related questions. The term correlation refers to the mutual relationship between two variables (X, Y). Therefore, there are usually two ways to understand the correlation between these two variables. One is to draw a scatter plot, and the other is to calculate the correlation coefficient (that is, to express the correlation in terms of the strength of the degree of correlation, the quantity of similarities and differences in a relevant direction). Correlation analysis attempts to use the correlation coefficient to measure the relationship between two variables. Correlation analysis cannot infer anything, it can only point out the high or low degree of correlation between two variables. Cross analysis is to compare the data of two or more groups, such as correlation analysis, chi-square test, T test, regression analysis and so on (Wu, 2001). In this study, the SPSS software was used in the cross analysis of the subjects.

2. Descriptive statistics

Explanatory data analysis (also referred to as initial data analysis) within descriptive statistics is the most basic and very important step in data analysis, and the success or failure of data analysis is often determined during this step. Descriptive statistics include the basic characteristics of data, such as: mean, standard deviation, proportion (chart), etc. The general classification methods are: the number of central trends and the number of differences (Yue, 2012a).

(1) Basic data analysis: Data error detection, data input errors, and finding possible outliers. Preliminary exploration of data characteristics, such as data concentration and distribution trends. Verify that the known result is the same as the known result (Yue, 2012b).
(2) Measures of central tendency: A. Average (Mean; expected value), arithmetic mean (Average); Weighted Average; others (geometric mean, blending mean). B. Median: Half of the values are larger than the median, and half of the values are smaller than the median. C. Mode: the value with the most occurrences (Yue, 2012b).
(3) Quantity variance: A. Range: the difference between the maximum and minimum values (Range = Max-Min). B. Quartile Deviation: Quartile (Quartile; Q1), 3/4 of the values are larger than Q1, and 1/4 of the values are smaller than Q1; interquartile difference = Q3–Q1.
(4) Variance (Variance; $\sigma 2$) and standard deviation (Standard Deviation; σ) (Yue, 2004). This study uses descriptive statistical analysis, which can be learned

literally, and makes statistical comparisons of concise and easy-to-understand data, such as percentages or bar graphs, etc.

3. Factor analysis

A century ago, when conducting some of the world's first intelligence-based tests, researchers noticed an interesting phenomenon: Suppose someone has taken two different cognitive ability tests, A and B, if he has a good score on test A, then he often does well on test B; conversely, if he does not do well on test A, he often does not do well on test B. Such correlations seem to imply that these different cognitive ability tests, in addition to testing specific cognitive abilities, can also measure common elements of a certain cognitive ability. Factor analysis is a statistical technique used to analyze such common elements. In the early 1900s, British psychologist Charles Spearman proposed the method of factor analysis; he used the correlation between these test results to calculate a single factor that could represent the common part of all tests, and called it the general factor, which could also be shortened to G (Hung & Liang, 2016).

The purpose of factor analysis is to explain what (if any) pattern relationship exists between variables. It can even be said that it attempts to know whether or not "we can use a small number of latent variables called factors to explain the relationship between most observable variables." Most of today's statistical methods are used to study the relationship between independent variables and dependent variables. However, factor analysis is different, it is used to study how multiple variables are related, in order to find out the latent variables which influence these variables. These inferred latent variables are called factors. A typical factor analysis mainly answers the following questions: (1) Several different factors are needed to explain the correlation between these observed variables. (2) What the nature of these factors are. (3) Whether these hypothesized factors can reasonably and completely explain the observed data. (4) What unique factors are involved in each observed variable. The main purpose of factor analysis is to test the "validity" of the questionnaire. In other words, the most commonly used method to test the validity of the questionnaire is "factor analysis." If we use "factor analysis" to test the test tool (i.e. the questionnaire) and successfully extract a common factor, one which is very close to the psychological characteristics of the theoretical framework, it means that the questionnaire (test) has "construction validity (Hung & Liang, 2016)."

(1) Reliability: Refers to the reliability of an indicator. Both measure dependability, consistency and stability. Therefore, reliability analysis is to analyze the reliability of each questionnaire result obtained after the questionnaire survey, and most of them use the "Cronbach α" coefficient method to analyze whether the results after the questionnaire survey are reliable (Bryman & Cramer, 1997). If the Cronbach alpha value is above 0.8, it indicates that the questionnaire results have a high reliability (Wu, 2001).
(2) Validity: Validity represents the accuracy of the results. Using the example of shooting at a target, if five shots are fired, the five shots are all clustered very close, but not close to the bull's-eye. Then the shots could be considered reliable,

but not valid. If the five shots are all close to the bullseye, but very scattered, then the shots could be considered valid, but not reliable. Three characteristics of validity: A. "Validity" refers to the correctness or reliability of the questionnaire results. B. "Validity" refers to the appropriateness and representativeness of the content of the test or questionnaire or the questions of the questionnaire. That is, the content of the questionnaire can reflect the psychological characteristics to be studied, and whether it can achieve the purpose or behavioral constructs to be tested. C. "Validity" refers to the degree to which a test can measure a theoretical trait or concept, that is, how much the actual test score can explain a certain psychological trait (i.e., customer satisfaction and loyalty). If it has good validity, it means that it can effectively explain the psychological characteristics of the subjects (questionnaire respondents). The three characteristics listed above, combined, can help explain the concept of validity. We can say that validity analysis is used to analyze whether the results of questionnaires can reflect the reliability of the final purpose we are discussing, whether it is representative, and the degree to which the questionnaire is very suitable for the final traits we are discussing (Wu, 2001).

(3) Kaiser–Meyer–Olkin (KMO) measure of sampling adequacy: According to Kaiser (1974), the research aspect or variable can be judged from the value of the Kaiser–Meyer–Olkin Measure of Sampling Adequacy to determine whether the item is suitable for factor analysis. The criteria for its judgment are as follows, according to the suitability of factor analysis of KMO statistic value: A, above 0.90 is very suitable for factor analysis B, above 0.80 is suitable for factor analysis C, above 0.70 can be used for factor analysis D, above 0.60 is barely suitable for factor analysis E, above 0.50 is not suitable for factor analysis F, below 0.50 is very unsuitable for factor analysis (Kaiser, 1974).

(4) Factor rotation methods: The principle of rotation is designed to conform to Thurstone's (1947) "Simple Structure," that is, to make each variable in the factor matrix after rotation be attributed to only one or a few factors. The goal of rotation is to increase the load of factors that are zero or close to zero in the matrix to reduce the complexity of the factors and make the explanation of the factors from complicated to simple. Usually there are four methods of orthogonal rotation: A, Varimax; B, Biquartimax; C, Quartimax; D, Equamax. Each method of orthogonal rotation is further divided into two variations of "Raw" and "Normalized," so there are a total of eight choices. As for the meaning of each rotation, we don't feel the need to further elaborate (Overall & Klett, 1972).

This study uses factor analysis in an attempt to explain satisfaction regarding the final developed cultural creative products and the underlying relationships that affect these variables.

4.5.2 Questionnaire

(1) Questionnaire title: The feasibility of using the conceptual form of the local cultural and historical design symbol knowledge base, as applied to the design of cultural creative products.
(2) Subjects: Those who have experience in product design are selected as the subjects to be tested. The "foreign designers" are students of the Department of Industrial Design of Tunghai University. The purpose for choosing this group of test subjects is because the goal of the knowledge base is for it to be used by non-local designers, these results will be used for follow-up inspection and analysis of the knowledge base's extensive user base. In addition, "local designers" were chosen from the product development and design department of the Cao Xie Dun Local Culture and Education Association. The reason is that the products developed by the Cao Xie Dun Association are mainly for the sale and consumption of cultural creative products in the tourist area of Caotun Township. Thus, relevant members of the Cao Xie Dun Association were chosen to conduct the inspection of final product designs (Table 4.12).
(3) Time and place of the questionnaire: Students from the Department of Industrial Design of Tunghai University, who have been trained for two months, were located in the professional classroom on the 2F of the Engineering Building of Tunghai University. In addition, members of the Cao Xie Dun local cultural and

Table 4.12 Specialties of respondents

Specific subject identity	Population	Proportion (%)	Professional background
Students from the Department of Industrial Design of Tunghai University (Hereinafter referred to as the University Department)	35	77.8	Industrial Product Design, Human Factors and Ergonomics Design, Market Analysis, Material Application and Research, Computer 3D Rendering and Modeling Technology
Members of the Cao Xie Dun local cultural and educational association (hereinafter referred to as the Cao Xie Dun association)	10	22.2	Technicians at the National Taiwan Craft Research and Development Institute, Community Cultural Guides, Artisans Department within the Grassroots Alliance, local dialects, Practical cultural and creative Product Development within the Grassroots Alliance, archaeological studies, education of straw patchwork dyeing and weaving, community building, financial management, field studies

Table 4.13 Training duration and locations of respondents

Identity	Training period	Training location
Students from the Department of Industrial Design of Tunghai University (foreign designers)	Two months	Tunghai University Engineering Building 2F Professional Classroom
Members of the Cao Xie Dun local cultural and educational association (local designers)	Two months	Caotun Straw Craft Culture Center 1F Workshop

educational association have been trained for two months at the workshop on the 1F of the straw craft culture center in Caotun (Table 4.13).

(4) Purpose of the questionnaire: To apply the theoretical framework of semiotics, to standardize the conceptual form of the knowledge base of Caotun local cultural and historical design symbols, and to assist the feasibility analysis of the design and use of local cultural creative products.

(5) Close ended questionnaire: (1) Those who have received specific training in design, specifically to develop products, such as college students (foreign designers) and members of the Cao Xie Dun Association (local designers). (2) 45 total valid questionnaires. (3) Questionnaire location: Industrial Design Professional Classroom of Tunghai University, Cao-tun Straw Craft Culture Center.

(6) Analysis focus: (1) Appropriate time training period for effective usage of the knowledge base. (2) Difference analysis before and after the use of the knowledge base. (3) Comparative analysis of difficulty and ease of development of cultural creative product development models (4) Satisfaction analysis of cultural product achievements after development.

4.5.3 Analysis of the Basic Data of the Subjects

A total of 45 valid questionnaires were conducted in this survey. There were 35 s-year students of Tunghai University and 10 members of the Cao Xie Dun Cultural and Educational Association. The former were professionally trained in design and education, and the latter were professionally trained in technology and development. The questionnaires were conducted based on these two personalities. A total of 45 valid questionnaires were collected, and the recovery rate of valid questionnaires was 100%. In terms of gender distributions, those who identified as "male" accounted for 24 people (46.7%), and "female" accounted for 21 people (35.3%). Age is divided into "under 29," which accounted for 35 people (77.8%). "Over 30 years old" 10 people (22.2%). Eight people (17.8%) have an education level of "below high school or vocational school," while 37 people (82.2%) selected the education level of "college

Table 4.14 Cross tabulation analysis: respondents' occupations and basic information

Option		Identity					
		The University department		The Cao Xie Dun association		Total	
		Number of people	Proportion (%)	Number of people	Proportion (%)	Number of people	Proportion (%)
Gender	Male	15	42.9	9	90.0	24	46.7
	Female	20	57.1	1	10.0	21	35.3
Age	Under 29	35	100	0	0	35	77.8
	Over 30 years old	0	0	10	100	10	22.2
An education level	Below high school or vocational school	0	0	8	80.0	8	17.8
	College or above	35	100	2	20.0	37	82.2
In terms of current employment	Students	35	100	0	0	35	77.8
	Member of Society	0	0	10	100	10	22.2

or above." In terms of current employment, 35 (77.8%) respondents selected "Students," all of whom are university students; while 10 (22.2%) respondents selected "Member of Society," all of whom are members of the Cao Xie Dun Association (Table 4.14).

4.5.4 Knowledge Base Training Appropriate Time Analysis

1. How long does it take you to become comfortably familiar with the knowledge base?

With the availability of the statistical results of the two types of subjects, 17 people (38%) reported that they trained for "one month." The data shows that respondents do not need two months, one month's time was all most people required to proficiently implement the development model. The options "one to two days" and "one week" each accounted for 9 people (20%) and ranked second (Fig. 4.13 and Table 4.15).

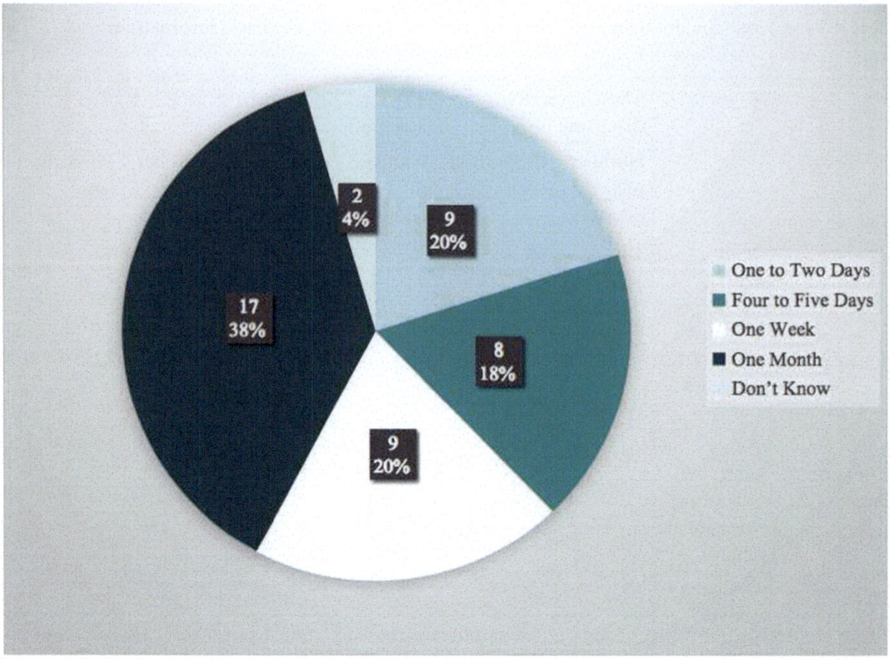

Fig. 4.13 Knowledge base training time percentage analysis

4.5.5 Feasibility Analysis of the Conceptual Form of the Knowledge Base of Cultural and Historical Design Symbols to Assist Designers in Developing Products

1. After using the knowledge base, do you think there is a way to develop cultural creative products?

The data shows that after using the knowledge base, cultural creative products can be effectively designed and developed. With 45 people (100%) choosing "Yes." While "No" accounts for 0 people (0%), which proves that the knowledge base can assist designers in developing products as it received high evaluations during test runs (Fig. 4.14 and Table 4.16).

1. If the content of the knowledge base of cultural and historical symbols in Caotun are integrated into other cases of local industries, can the design and development effect be brought into play?

The following statistics display the response towards the question of whether or not introducing the Caotun cultural and historical symbol knowledge base into the industries of other localities can produce the same beneficial design and development

Table 4.15 Cross tabulation analysis: duration of training on design thinking process

Identity	Option											
	One to two days		Four to five days		One week		One month		Don't know		Total	
	Number of people	Proportion (%)	Number of people	Proportion (%)	Number of people	Proportion (%)	Number of people	Proportion (%)	Number of people	Proportion (%)	Population	Proportion (%)
The University department	6	13	5	11.1	8	18	14	31	2	4	35	78
The Cao Xie Dun association	3	7	3	7	1	2	3	7	0	0	10	22
Total	9	20	8	18	9	20	17	38	2	4	45	100

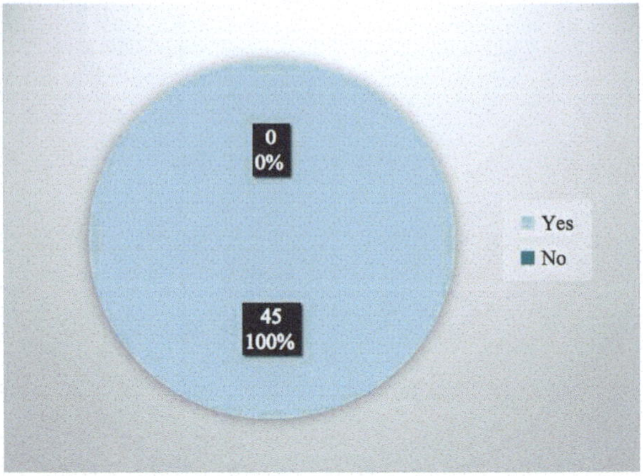

Fig. 4.14 Percentage analysis of whether cultural creative products can be effectively designed and developed after using the knowledge base

Table 4.16 Feasibility analysis on design model presented in this study

Question title	Option					
	Yes		No		Total	
	Number of people	Proportion (%)	Number of people	Proportion (%)	Number of people	Proportion (%)
Can you effectively design and develop cultural creative products after using the knowledge base?	45	100	0	0	45	100.0

effect. In response, 40 people chose "Yes," accounting for the majority (89%). 5 people (11%) chose the option "No." These results show that it is feasible to apply the same framework to an alternate location to achieve the same benefits in various industries in the production of cultural creative products. However, there are a small number of 5 people in the data who think that it cannot be used in the development of cultural creative products in various other places. These responses can be used as a starting point for future discussions regarding this matter (Fig. 4.15 and Table 4.17).

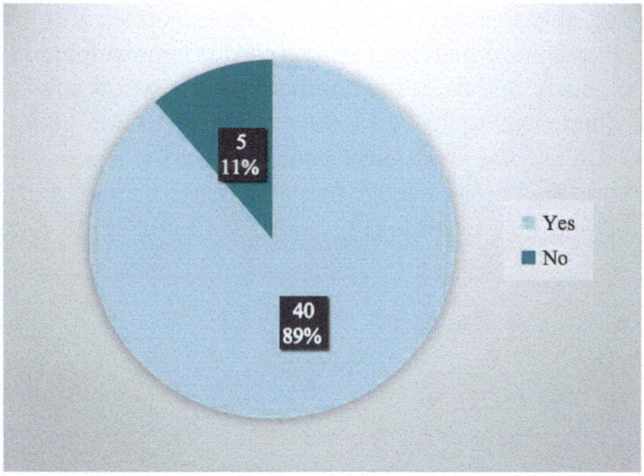

Fig. 4.15 Feasibility percentage analysis of applying the knowledge base structure in different places

Table 4.17 Feasibility analysis of applying design model used in this study in other cases

Question title	Option					
	Yes		No		Total	
	Number of people	Proportion (%)	Number of people	Proportion (%)	Number of people	Proportion (%)
Can the case of the transformation of Caotun cultural and historical symbolic knowledge base model into other local industries play a positive role in development?	40	89	5	11	45	100.0

4.5.6 *Analysis of Differences Before and After Using the Knowledge Base*

There are 35 respondents who are from the University Department and 10 respondents who are from the Cao Xie Dun Association, for a total of 45 people. It can be seen that there is a clear difference before and after using the knowledge base among members of the University Department(foreign designers) and the Cao Xie Dun Association (local designers). "Before using the knowledge base," 35 people (78%) proved to be able to use it, and 10 people (22%) did not know how to use it. "After using the knowledge base," the number of respondents who proved proficient at being able to use the knowledge base increased to 44, accounting for 98% of total respondents, which is close to 100%. When compared to the results from before using the knowledge base, the number of users who know how to use the knowledge base has increased to 20%; Those who still have not displayed a strong understanding of how to use the knowledge base account for 1 person (2%). These results prove that after using the knowledge base, the implementation of the local product development model is very feasible (Figs. 4.16, 4.17 and Table 4.18).

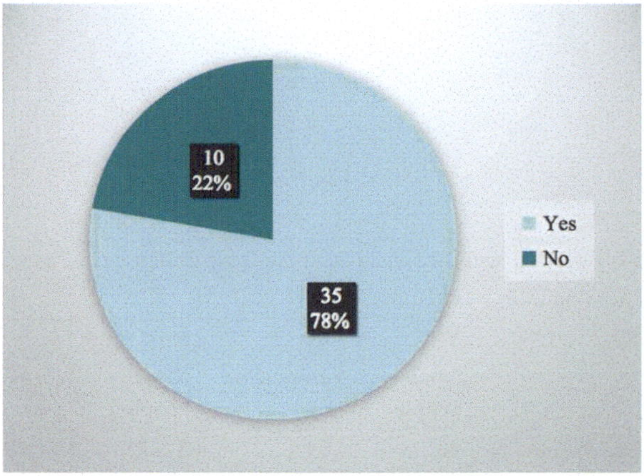

Fig. 4.16 Before using the knowledge base

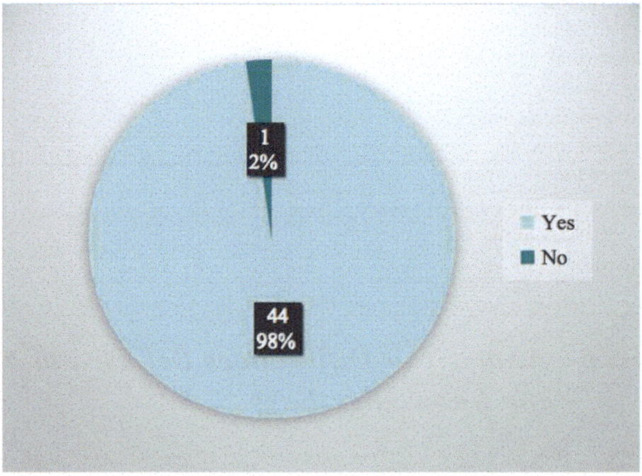

Fig. 4.17 After using the knowledge base

Table 4.18 Statistical analysis on effectiveness of design model: before and after

Question title		Difference before and after (expressed as a percentage)						
		Yes		No		Total		Difference before and after (expressed as a percentage)
		Population	Proportion (%)	Population	Proportion (%)	Population	Proportion (%)	Population
This is the first time you see the content of the knowledge base of local cultural and historical symbols, can you understand how to use it?	Before using the knowledge base	35	78	10	22	45	100	44
If the process method of using the knowledge base was explained in detail, would your understanding of house to implement it be more clear?	After using the knowledge base	44	98	1	2	45	100	56

4.5.7 *Analysis and Comparison of the Ease or Difficulty of the Development Mode of Cultural Creative Products*

1. Which stage of the cultural creative product design process do you believe is the hardest?

There is a clear gap between the selection of the University Department and the Cao Xie Dun Association on this question; 23% of the University Department believed the most difficult stage of the process was "extract [ing] the shape, color, material, texture and function information from the knowledge base." In these responses, it was learned that the university students indicated that there is room for improvement regarding extracting information from the knowledge base. For example, the knowledge base can be used with more levels, convenience, simplicity and become easier to understand. Cao Xie Dun Association chose "Brainstorming/Idea Sketch" as the most difficult stage, accounting for 38% of total responses. It can be seen that local residents have not been trained professionally in the relevant fields of design, and thus creative concepts are less likely to be conceived and realized (Fig. 4.18 and Table 4.19).

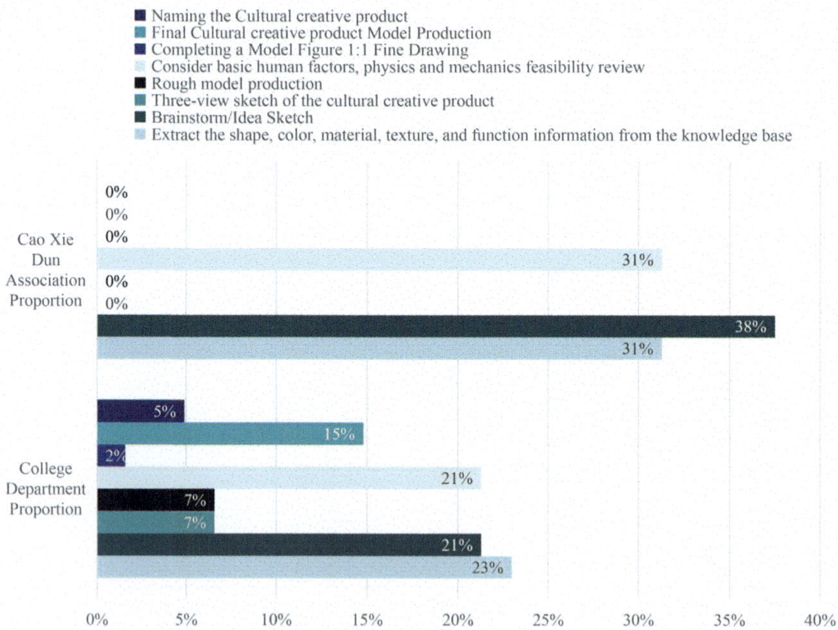

Fig. 4.18 Bar chart of the most difficult steps in the design process of cultural creative products

Table 4.19 Most difficult stage of design thinking process

Question title	Option	College department proportion (expressed as a percentage)	Cao Xie Dun association proportion (expressed as a percentage)
Which stage of the cultural creative product design process is the most difficult?	Extract the shape, color, material, texture, and function information from the knowledge base	23%	31%
	Brainstorm/Idea sketch	21%	38%
	Three-view sketch of the cultural creative product	7%	0%
	Rough model production	7%	0%
	Consider basic human factors, physics and mechanics feasibility review	21%	31%
	Completing a model figure 1:1 fine drawing	2%	0%
	Final cultural creative product model production	15%	0%
	Naming the cultural creative product	5%	0%
Total number of people		35	10

1. What do you think is the easiest step in the design process of cultural creative products?

For the university department, the option of "rough model production" accounted for 23% of total respondents. For the university department, the materials are easy to obtain and can be used to the fullest. Another option in the university department, "Braining Idea Sketch," was ranked second with 19% of total responses choosing this step as the easiest in the process, which can be explained by the students of the University Department having been professionally trained at the university level in the field of design. The Cao Xie Dun Association accounted for 29% of the option "Final Cultural creative product Model Production." This is likely explained by local residents having considerable experience in the technology and materials for making the final model (Fig. 4.19 and Table 4.20).

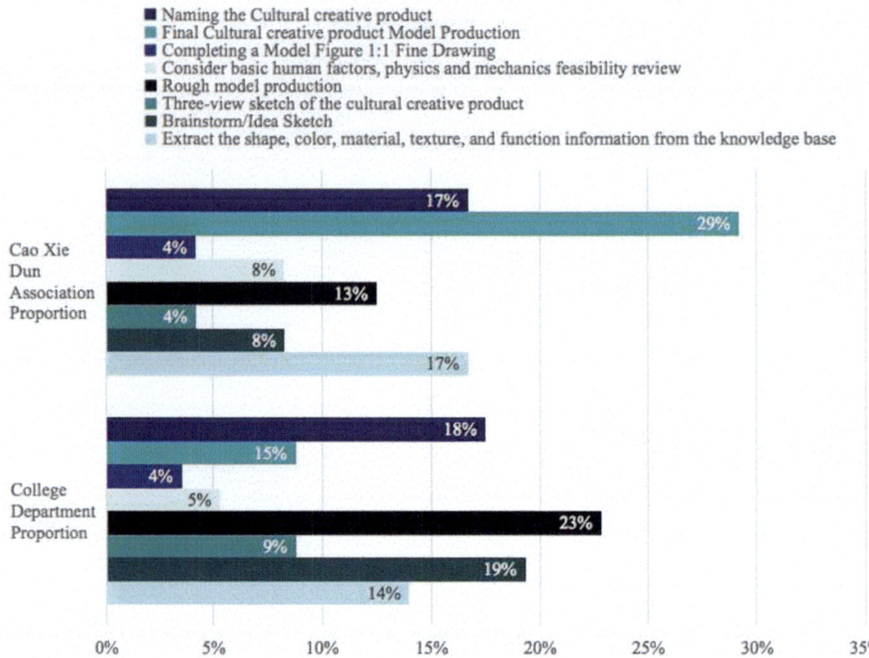

Fig. 4.19 Easiest step in the cultural creative product design process bar chart

4.5.8 Analysis of Satisfaction of Cultural Creative Product Achievements After Design

The analysis of the overall satisfaction factors of cultural creative products designed using the local cultural and historical symbol knowledge base is as follows:

(1) Factor 1: The options include color satisfaction, design satisfaction, texture satisfaction, overall satisfaction, and others. This factor is named "cultural creative product cultural concreteness." The data proves that the subjects give this factor—the shape, color, texture, and overallity of cultural creative products, the highest priority. Further showing the subjects have the highest satisfaction with the tangible cultural elements of local products. The average values of "Satisfaction of Shape" and "Satisfaction of Overall Sense" are (3.40), which is the highest in factor 1, and the satisfaction degree of shape design and overall cultural creative products after design is high. The average value of "Texture Satisfaction" (3.20) is relatively low, because most of the respondents are concerned with the production of products, and the degree of refinement will directly affect the texture, which is related to the designer's technology and materials.
(2) Factor 2: The options include "uniqueness," "symbolism," "historicity" and "story-like" and other satisfactions. This factor is named "culture creative

Table 4.20 Easiest stage of design thinking process

Question title	Option	College department proportion (expressed as a percentage)	Cao Xie Dun association proportion (expressed as a percentage)
Which step do you feel is the easiest in the cultural creative product design process?	Extract the shape, color, material, texture, and function information from the knowledge base	14%	17%
	Brainstorm/Idea sketch	19%	8%
	Three-view sketch of the cultural creative product	9%	4%
	Rough model production	23%	13%
	Consider basic human factors, physics and mechanics feasibility review	5%	8%
	Completing a model figure 1:1 fine drawing	4%	4%
	Final cultural creative product model production	9%	29%
	Naming the cultural creative product	18%	17%
Total number of people		35	10

commodity cultural connotation," according to the responses regarding the second factor, the cultural significance of local cultural creative products takes second place. Among the options, the average response for "symbolism" (3.51) is the highest data in the second factor, which means that products developed with the knowledge base can condense the special cultural connotation of the place, convey the symbol of the place in the product, and clearly identify the local code.

(3) Factor 3: The option is "Material Satisfaction"; this factor is called "Cultural creative products Material Application." The average value of the option "Material Satisfaction" is the highest among factors 1–4 (3.56), which proves that cultural creative products are suitable for use within local industrial resources such as straw and pottery materials. The results also indicate that materials can add value to the unique essence of each locality and their cultural creative products.

(4) Factor 4: The "functional satisfaction" factor is named "Cultural creative products Functional Use." The average score of respondents' satisfaction with the results of cultural creative products after development is low (3.07). It shows that the respondents tend to be practical in the design functions of cultural creative products, and the designed products do not have much functionality and are more inclined to decoration, so the reason why the average of the total evaluations is the lowest (Table 4.21).

Table 4.21 Satisfactory levels and factor analysis of design results

Statistical Methods / Option	Descriptive Analysis		After Factor Analysis			
	Average Evaluation	Variance	Factor1	Factor2	Factor3	Factor4
Color Satisfaction	3.33	.74	.878	.103	-.034	-.232
Styling Satisfaction	3.40	.84	.688	.203	.223	.367
Texture Satisfaction	3.20	.73	.679	.195	.526	.194
Overall Satisfaction	3.40	.84	.628	.538	.255	-.001
Symbolic Satisfaction	3.51	.82	.196	.873	-.113	-.034
Uniqueness Satisfaction	3.22	.82	.110	.729	.529	.132
Historical Satisfaction	3.36	.65	.301	.558	.386	-.201
Storylike Satisfaction	3.42	.72	.074	.440	.492	-1.217
Material Satisfaction	3.56	.76	.160	.029	.874	-.056
Functional Satisfaction	3.07	.72	.018	-.077	-.098	.905
Factor Naming			Cultural creative product Cultural Concreteness	Cultural creative product Cultural Connotation	Cultural creative product Material Application	Cultural creative products Functional Use
Cumulative Total			23.287	46.526	64.582	75.264
Reliability .8165						

References

Andersen, A. (2000). *The First Book on Knowledge Management*. Business Weekly Publications Inc.

Birkhoff, G. (1973). *A Source Book of Classical Analysis*. Harvard university Press.

Berger, A. A. (1992). *Reading Matter: Multidisciplinary Perspectives on Material Culture*. Transaction Pub.

Berger, A. A. (1933). *Media Analysis Techniques*. Sage.

Bryman, A.., & Cramer, D. (1997). *Quantitative data analysis with SPSS for windows: A guide for social scientists*. London: Routledge.

Compiled by Caotun Town Office. (1986). *Caotun township chronicle*. The Caotun Township Office, Nantou County.

Chang, S. Y., Tsao, J., & Ku, L. (2009). A study on subconscious symbolic sign in salvador Dali's works. *Collection of Art Papers, 12*, 95–118.

Eco, U. (1984). *Semiotics and the Philosophy of Language*. Indiana University Press.

Fiske, J. (1989). *Understanding Popular Culture*. Unwin Hyman.

Fiske, J. (2010). *Introduction to Communication Studies* (3rd ed.). Routledge.

Hu, Y. H. (2010). On the proofs of the boundedness of the continuous function in the closed interval with the real number completeness. *Journal of Lishui University, 32*(5), 8–10. https://doi.org/10.3969/j.issn.1008-6749.2010.05.003.

Hiroshi, O. (1967). *The Foundations of Design*. Dongdu Study.

Hung, H. T., & Liang, S. R. (2016). An application of DEA to evaluate risk priority of the failure modes in FMEA. *Journal of Crisis Management, 13*(1), 1–8. https://doi.org/10.6459/JCM.201603_13(1).0001.

Hoshino, K. (1991). *Consumption in a Symbolic Society*. Yuan-Liou Publishing Co., Ltd.

Kaiser, H. F. (1974). An index of factorial simplicity. *Psychometrika, 39*, 31–36.

Lv, Q. F. (1984). *Principles of Shape*. Hsiung Shih Art Books Co., Ltd.

Lin, C. H. (2001). *Design Theory and Value—A Discussion of Design Culture and New Thinking*. Garden City.

Lin, C. H. (2014). *Design Fundamentals: Creative Thinking About Shape and Composition*. Chuan Hwa Publishing Ltd.

Lee, H. H. (2002). *The sign effects in the affective and communicattional aspects-using chair design as examples*. Master's thesis. Department of Design, National Taiwan Normal University, Taipei.

Liu, L. W. (2007). *Examples of Works for Practical Application of Product Design*. Sungood Books.

Mirzoeff, N. (1999). *An Introduction to Visual Culture*. Psychology Press.

Ng, K. C. (2018). Compound ideographs: Metaphor and metonymy as two poles. *Sun Yat-Sen Journal of Humanities, 45*, 21–45.

Overall, J. E., & Klett, C. J. (1972). *Applied Multivariate Analysis*. McGrawHill.

Piaget, J. (1970). *Structuralism*. Harper & Row.

Sallie, M. F. (1982). *Metaphorical Theology: Models of God in Religious Language*. Fortress Press.

Su, W. C., Yen, J., & Lee, C. F. (2007). Application of semiotics and cognitive psychology to visual design. *Journal of Humanities and Social Sciences, 3*(1), 95–104.

Sunkiss. (2003). *A study on evolution of telephone model*. Master's thesis. Graduate School of Applied Design, Shu-Te University, Kaohsiung.

Thurstone, L. L. (1947). *Multiple Factor Analysis: A Development and Expansion of Vectors of the Mind*. University of Chicago Press.

Wang, M. F. (2007). *The research of symbolism in local culture commodity developing mode—A case study in Nantou County*. Master's thesis. Department of Industrial Design, Tunghai University, Taichung.

Wang, M. F. (2020). *Cultural Realism and Virtualism Design Model*. Springer.

Weng, X. D., & Miyazaki, K. (1996). *The flower of the heart—Concepts and examples of the overall building of Japanese communities*. Caotun Township, National Taiwan Craft Research Institute.

Wu, M. L. (2001). *Introduction to Educational Action Research—Theory and Practice*. Wu-Nan Book Inc.

Wu, Z. F., Lan, T. X., & Luo, Z. Y. (2010). The study on system design requirements for knowledge management. *Management Sciences Research, 7*(1), 35–49.
Yang, Q. T. (1997). Composition (Vol. 1). San Min Book.
Yang, Y. F. (1998). *Cultural Foundation of Design.* Ai-Tai Publishing Co.
Yu, H. Y. (2001). Applied semiotics study on the signification modes between meanings and objects in the design process-with objects transferred from function to meanings of living as example. Master's thesis. Department of Industrial Design, National Yunlin University of Science and Technology, Yunlin.
Yue, C. J. (2012a). Mortality compression and longevity risk. *North American Actuarial Journal, 16*(4), 1–15.
Yue, C. J. (2012b). Mortality compression and longevity limits. *Contingencies,* (1), 42–45.
Yue, C. J. (2004). A note on optimal strategies in a generalized two-stage Bandit problem. *Communications in Statistics: Theory and Methods, 33*(7), 1577–1585.
You, K. Q. (2001). *Knowledge Management and Innovation.* Commonwealth Publishing Co., Ltd.

Chapter 5
Cultural and Creative Products Exhibition and Its Analysis Based on Scenario-Oriented Design

This book uses the conceptual form of the knowledge base of cultural and historical design symbols to shape local cultural creative products. After quantitative verification by non-local designers and local designers, it can indeed effectively solve the difficulties faced by designers in the design process of local cultural creative products. However, when consumers are unable to closely evaluate cultural creative products, it cannot be known if there is a difference in cultural understanding towards the cultural creative products developed using the knowledge base, or if there is a difference in purchase behavior as a result. Therefore, the scenario-based design method is used to plan a cultural creative product exhibition to evaluate the preference of foreign tourists and local residents towards culture creative commodities.

5.1 Literature on Scenario-Oriented Design

Artificial intelligence scientist Roger C. Shank explained in his book Tell Me A Story that human memory stores thousands of scripts, these scripts are the accumulated wisdom of our experiences in dealing with life events. When faced with particular circumstances or environments in life, we retrieve the corresponding scripts in response. If we combine these experiences with the corresponding scripts we use, they become stories. Shank believes that the use of scripts is a kind of wisdom human beings use to solve problems in daily life. Using scripts reduces the burden of thinking, and using correct scripts reduces risks and improves adaptability in various situations. If solving complex mathematical problems is an example of human intelligence, making good use of scripts would be an example of human wisdom (Yu et al., 2000). The method of storytelling using pictures, uses the method of telling stories to express and describe the theme of design. The focus of creation is not on practicality, but on narrating and expressing a concept or event. In the design and development stage, the method of telling a story by looking at pictures is mainly a research method that uses images to quickly collect stories from user experience

M.-F. Wang, *Creative Product Design With Cultural Codes*,
https://doi.org/10.1007/978-981-19-6767-2_5

and make stories present images. The first key is to obtain the influencing factors from the experience of consumers, and then combine the factors obtained at this stage with our environmental images to become the starting point for the designer's thought process when designing products (Tseng, 2002). Through viewing exhibitions, following roadmaps, telling stories and, learning about cultural exhibition planning, the public's understanding of culture can be deepened (Kuo, 2016). Local industries have also accumulated a lot of cultural resources. This snapshot of history is part of three main relationships of characters, the environment, and objects in the scenario-based design method. When these three relationships are combined with local events that resonate with each other, they are connected to form the content of a complete story, thereby approaching the public's perception and understanding of culture.

1. The relationship between characters, objects, and scenario

Storytelling is a form of basic human intelligence, and oral storytelling is a way through which we are able to describe the context of events. Listening to stories is another kind of wisdom. People often try to reproduce the original appearance of events and identify the meaning of the messages behind them. The basic structure of a "story" includes "characters, scenario, object-activity," through which the event is expressed in words or dictation, and people receive information through reading or the oral dictation of others, and reproduce the three aspects of "characters, scenario, object-activity" in their minds (Moggridge, 1993).

The situation simulated by the designer can be regarded as a virtual activity, which consists of the relationship between characters, scenario, and objects. "Characters, scenario, and object" refer to three groups of information, such as the characteristics and habits of people, the environmental conditions of the background (setting), and the features and functions of objects (Yu et al., 2000). For example, the ancestors of the local residents of Caotun represent "human knowledge"; the necessities of people's livelihood at that time referred to "the appearance of goods"; through the past and present industrial "situation" of Caotun, the cultural background of the Qing Dynasty was developed. The ancestors' stories regarding their trips to and from Lugang Danpushe formed the basis of cultural activities (Fig. 5.1).

(1) Script description of "Character"

Establishing a cast list is the basic action to define a "character." To write a story, of course, a protagonist is required. The script method also defines a "cast list" that usually lists three to five characters, representing the target group designed by the planner. The story structure will be organized through the "cast list" to transform and test the design goals, so the attributes of the actors cover a wide range, so as not to be too monotonous and rigid, miss different use plots, and further result in the development of flawed designs (Yu et al., 2000).

For the performance and exhibition of cultural creative products in this book, we must first draw up the background of the "Character Table," which is selected from the Lugang Danpushe during the Jiaqing period of the Qing Dynasty and the Japanese occupation period. The character table will have extensive coverage of design goals

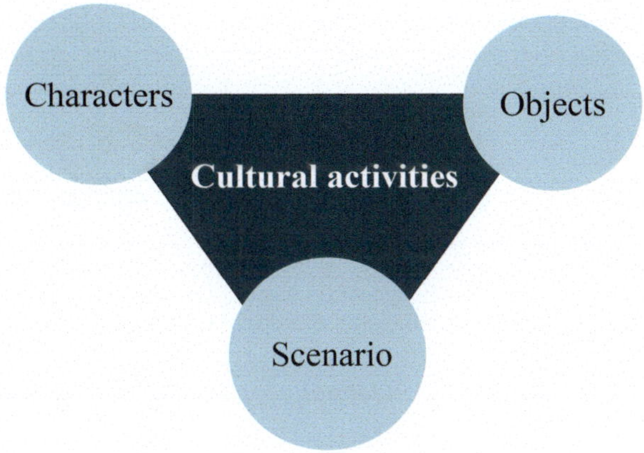

Fig. 5.1 The relationship between characters, objects, and scenario

and user groups, considering the name, gender, age, use environment, education, job title, personality, commonly used tools and hobbies, along with other traits of each character. Basically, because of the different background of each actor's character, the character's scope of application, job position, and route through the story are all different, forming a storyline that piques one's interest and reflects the lifestyle of the time period portrayed in the exhibition (Table 5.1).

(2) Script descriptions of past and present "Scenario(s)"

Scenario is a way of answering how events began and ended. Its main purpose is to discover and simulate the assumptions of the user's thinking, such as: behavioral activities, possible responses, and feasible opportunities (Yu et al., 2000). Observing and analyzing various behaviors of human beings in daily situations can help us better understand human behavior patterns, and the results of observation and analysis will also be of positive help to human behavioral decision-making (Jarke et al., 1998; Tseng, 2002). A local cultural industry must look at the "past" and "present." The so-called "past" refers to the past history and culture of the place, and the "present" refers to the new attractions and industries of the place. The present and the past stand as testaments to the abundant resources of the place. This study sets the Caotun situation as the past, with the Lugang Danpushe in the Qing Dynasty, Jiaqing, and Japanese occupation period as the background. At the same time, with the present: the changes in the current Cao-tun, such as: attractions and industries, are all good materials for writing script situations.

(3) Script description for "Map"

After establishing the protagonist and the situation, at the same time construct the "map" type to reflect the scope of the script to explore, and frame the world we want to explore. There are two types of maps:

A. "Log Map," which captures daily life activities, such as a day, a week, a season or a year (Yu et al., 2000), as shown in Table 5.2.

Table 5.1 List of ancestral characters

Picture			
Name	Ancestors during the Qianglong and Jiaqing years-Bragging grandpa	Ancestors during the Qianglong and Jiaqing years-Fan Shu Sun	Ancestors during the Qianglong and Jiaqing years-Fan Shu Grandma
Use environment	Route 1 of Lugang Danpushe	Route 2 of Lugang Danpushe	Route 3 of Lugang Danpushe
Gender	Male	Male	Female
Age	64	19	60
Education	Uneducated	Uneducated	Uneducated
Job title	Sell food such as tea, snacks and meals in the tea shop	Selling household hardware, including cloth, salt, oil, gunpowder, iron and urns	Making straw sandals and selling straw sandals
Personality	Humorous, gritty and down-to-earth	Filial piety, diligence and hard work, Taiwanese Mandarin accent	Good wife and good mother, frugal and stingy
Commonly used tools	One root pole, three-layer wooden box, fan, hat, bamboo kettle, towel, straw sandals	One root pole, hat, wooden box, wearing straw sandals, towel, straw sandals	Hairpin, straw sandals, chair bar, straw sandal rake, straw sandal slot, straw sandal waist, straw mallet, scissors, straw
Interests	Brag while making tea	Imitating puppet characters	Watching puppet shows outdoors

Table 5.2 Map journal (Yu et al., 2000)

Family / Process	Grandfather	Grandmother	Father	Mother	Eldest Grandson	Eldest Daughter	Friend
Get up							
Breakfast							
Morning							
Before and After Lunch							
Afternoon							
Before and After Dinner							
Evening							
Retire for the night							

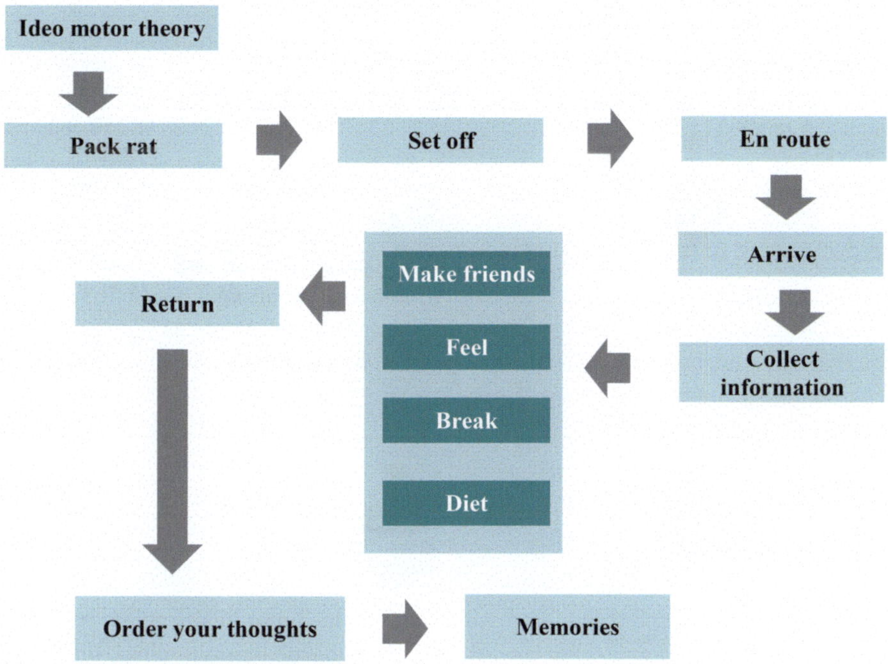

Fig. 5.2 Journey map (Yu et al., 2000)

B. A "journey map" describes a single life cycle from occurrence to end, from beginning to end. The event fragments in each stage of this cycle have their own correlation and sequence, like travel from beginning to end (Yu et al., 2000), as shown in Fig. 5.2.

Combining the above-mentioned "log map" and "journey map," the script descriptions of the two maps can be applied to the type of map for planning the roadmap which properly reflects in the cultural creative product exhibition in this study.

2. The road map of the ancestors of Caotun, the Qing Dynasty, Jiaqing and the Japanese occupation period. A route map of the ancestor's path from Lugang to Puli-she serves as a reference index from the occurrence to the end of the event (Compiled by Caotun Town Office, 1986, 2005) (Table 5.3).

5.2 Cultural and Creative Products Exhibition and Its Procedure Based on Scenario-Oriented Design

In order to clearly understand the process of scenario-based planning of the culture creative exhibition, the following four steps will be explained in detail:

Table 5.3 Routes taken by ancestors

Route	Ancestors	Route map	Route process content
Route 1 during the Qing dynasty			North from Dadun (Taichung), Ah Zhaowu (Wufeng), Cao Xie Dun (Caotun), Nei Muzha (Caotun Tucheng), towards Puli-she
Route 2 during the Jiaqing period			Lugang, Changhua, Kuaiguan, Shekou, Xianzhuang, Beitou, to Cao Xie Dun, east to Puli-she
Route 3 during the Japanese occupation			Changhua, Ding Chie Lao (Coatun Xinzhuang), Caotun (Fen Tsao Road), East to Puli

1. First, collect the relevant academic literature on the scenario-based design method, analyze and write the three elements of the script, such as the character table, the scenario and the map of the ancestors of Caotun.
2. The first step constitutes the three elements of the scenario-based cultural creative product exhibition, and the Lugang Danpushe is the main story to write a reasonable script that starts, inherits, transfers and integrates.
3. After the script of the Lugang Danpushe is drawn up, a life-size monopoly-style exhibit space will be created in coordination with the exhibition space. With Lugang set as "GO"(start point), Caotun is set as "free parking" (midway rest point), and Puli is set as the final tile on the game map. In total, there are ten total cultural creative product exhibit "stations," including the starting point in Lugang and the final point in Puli. Allowing participants and respondents to experience the display content by viewing the exhibit, tell stories, walk the map, and further learn about local culture.

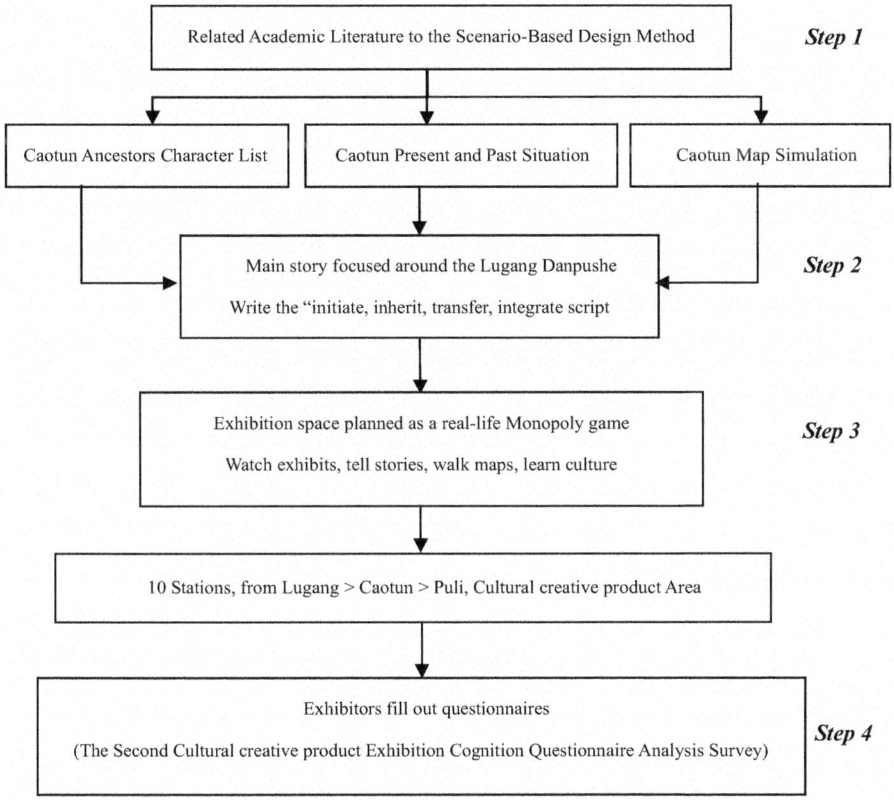

Fig. 5.3 Steps for the scenario-based design and planning process for the cultural creative product exhibition

4. For the three cultural creative product exhibitions, visitors will fill out questionnaires to analyze their cultural understanding, consumption behavior, and the efficacy of the scenario-based exhibition with respect to the Caotun local cultural creative product exhibition (Fig. 5.3).

5.3 Scenario Composition and Setting

In recent years, the scenario-based design method has been widely used by IDEO and related research by Makela and Mattelmaki (2002). These are all storytelling for a specific event, but if you follow the principles of storytelling, it is very important to start, inherit, transfer and integrate (Tseng, 2002). Exhibition display design and planning can be divided into the following three points: Visual Presentation, referred to as VP; Point of Sales Presentation, referred to as PP; Item Presentation, referred to as IP, the above three points are also known as the 3P of exhibition display design and planning (The Beautiful Home Editorial Department, 2019). Although control over the process of the scenario-based design method is abstract, it can still be summarized as a certain model for practice and operation by users. The creativity of the scenario-based design method gives it rhythm, such as the basic principles of our traditional composition "initiate, inherit, transfer, and integrate (Gabriel, 1998; Yu et al., 2000)." Among the five major industrial resources of Natural, Production, Human, Cultural and Landscape, "cultural" is the soul of each resource: The production of local "cultural" is the process by which people in the region use the resources of "natural" to create "landscape" and "production." These processes are the development of the region's history and culture, local life culture, and local customs and religious beliefs. Therefore, the generation of "culture" is a process of the application, interweaving and generation of local "human," "landscape," "natural," and "production (Huang, 2001)." Therefore, in the case of Caotun, the most recognized "Lugang Danpushe" will be used in the scenario-based design method, the planning story of the main cultural resources is the main axis.

5.3.1 "Initiate" Planning Steps

The "Initiate" part of the exhibit is interpreted through the analysis of the lifestyle of Caotun, the appearance of the ancestors, the script map, among other things, and guides visitors upon their entrance into the exhibit by providing a scenario that serves as a backdrop for the exhibit. The resources of natural, production, human, cultural and landscape in this book are mainly Caotun cultural resources. According to the Caotun Township Chronicle, it is recorded that travelers from Lugang would pass through Caotun on their trip to Puli (Compiled by Caotun Town Office, 1986, 2005). The story background and planning for the cultural creative product exhibition was decided based upon the purpose and route of Lugang Danpushe at that time:

1. Purpose: During the Qing dynasty, Jiaqing period, and the Japanese occupation, resources were scarce. For the livelihood of their families, the ancestors found time, in their regular schedule centered around farming, to travel to the Lugang port and carry goods through Caotun to Puli to buy and sell.
2. Three largest ports in Taiwan at the time: Yi Fu (Tainan Prefecture), Er Lu (Lugang), San Bangkah (Wanhua); Lugang was the largest shipping port in central Taiwan.
3. Route: The porter relies on both his shoulders to carry goods from Lukang, Changhua, Kuaiguan, Shekou, Xianzhuang, Beitou, and Cao Xie Dun, eventually entering Puli-she from the East, stopping in between to rest, spend the night, and replace their straw sandals (Fig. 5.4).

Visual Presentation, referred to as VP for short, usually chooses important exhibition and sales spaces that are popular and have high color contrast, which can attract customers' when they first turn their attention to the exhibit. Through the design of a strong visual theme, what the exhibition has to offer is conveyed to viewers

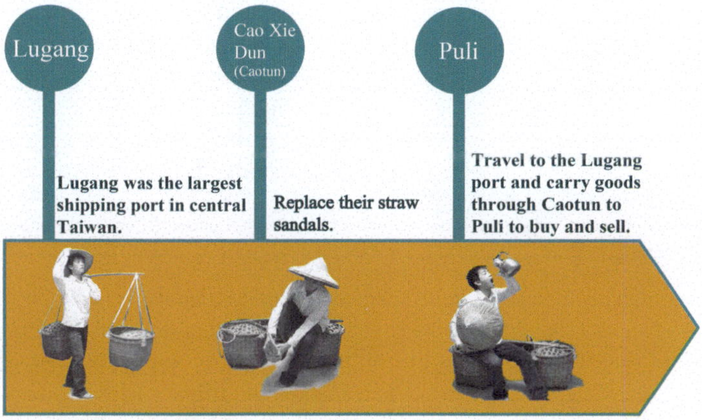

Fig. 5.4 The purpose of the Lugang Danpushe

and consumers. This can include important information and the display of thematically linked products (The Beautiful Home Editorial Department, 2019). The posters, DM promotions, and commemorative exhibition CDs at the cultural creative products exhibition all convey to the viewer important visual images of the local culture through a visual theme by design (Figs. 5.5, 5.6, 5.7, 5.8 and 5.9).

5.3.2 "Inherit" Planning Steps

Through the writing and reading experience of troublesome scripts and dreamy (ideal) scripts, the "inherit" style brings out (externalizes) abstract tacit knowledge such as usage situations, interaction modes, and key issues, and accumulates the tension of inspiration for solving problems (Green & Jordan, 2002; Yu et al., 2000). Point of Sales Presentation, referred to as PP, displays the exhibits, attracts the attention of viewers and consumers' desire to buy by means of contextual atmosphere and thematic focus, and quickly finds products which suit them (The Beautiful Home Editorial Department, 2019). Through the story of "Lugang Danpushe," the exhibition takes every stop made by the ancestors on their route and creates an engaging design of the exhibition space through the medium of a life-size monopoly game. Essentially, the exhibit is able to, under the context of a game of monopoly, convert the resources of natural, production, human, cultural and landscape in Caotun to make the culture creative commodity exhibition vivid. Not only strengthening the content of the exhibition story and the creation of a scenario within the minds of viewers, but also allow the public to get closer to and experience the cultural characteristics of the place. Thus helping achieve an understanding among tourists and local residents alike, who are then more able to understand the local characteristic cultural creative products (Fig. 5.10).

Fig. 5.5 Poster for the cultural creative product exhibition

5.3.3 "Transfer" Planning Steps

"Transfer" refers to the script used for application and the script used for interaction, turning inspiration into specific images such as application concepts, interactive concepts, and product simulations (Yu et al., 2000). The goal is to simulate the specific situations within each of the ten sites in the exhibition hall, and plan out the software and hardware content including location, product category, exhibition area and scenario which the viewers are supposed to experience (Table 5.4).

5.3.4 "Integration" Planning Steps

Finally, in addition to the promotion process of "initiate, inherit, transfer, and integrate," "Integration" uses the 5W1H of people, events, time, place, and objects as the basic writing mode of the drama score to fill in the script, which is also helpful to the participants. The situation is developed from the existing literature in a situation of nothingness (Yu et al., 2000). In this study, the scripts of "initiate, inherit, transfer, integrate" are connected in series, and a 1:20 scale model of the spatial circulation routes and the exhibition venue itself are simulated and planned on site.

Fig. 5.6 Introductive extension poster for each stop

Fig. 5.7 Cultural creative product exhibition invitation card front

Fig. 5.8 Cultural creative product exhibition invitation card back

1. Spatial circulation planning: In the three exhibition spaces, it was found that most of them were rectangular spaces. The first venue, Caotun Straw Craft Culture Museum, has a length of 1700 cm × 520 cm in width (Fig. 5.11); the second venue, Caotun Town Office 2F Exhibition Center, has a space of 405 square meters (Fig. 5.12); the third venue, Taichung City Dadun Cultural Center, Wenying Building, has a length of 1980 cm x width of 720 cm (Fig. 5.13).
2. 1:20 scale model of the exhibition site: A model of the size of the space at the exhibition site, with a scale of 1:20, planned to simulate the moving lines of 10 stations, table boards, booths, large dice and three station signs (Fig. 5.14a–c).

5.3.5 Scenario Planning of the Culture Creative Commodities Exhibition

Item Presentation, referred to as IP, uses showcases, shelves, cabinets and other props to display the actual product area in the exhibition hall or store. The exhibition of cultural creative products is presented statically (The Beautiful Home Editorial Department, 2019). In addition to the explanation provided by the exhibition itself, in order for the full benefits of the scenario-based design method to be realized, the exhibition scene should be able to achieve a strong emotional connection with the audience. In the production of the scene, it is necessary to design hardware

Fig. 5.9 Cultural creative product exhibition CD cover

dashboards, stop signs and properly place the tools and artifacts once used by the ancestors, along with various software components, including guide instructions and Taiwanese folk music, etc., to increase the atmosphere of the exhibition and create a stronger connection with the audience (Table 5.5).

5.4 Space Planning of Cultural and Creative Product Exhibitions

Following the steps of "initiate, inherit, transfer, and integrate," the main storyline for the script was established and written. The script was focused around the story "Lugang Danpushe," which was set as the main scene. The stations one through ten were planned out, following the route traveled by the ancestors. Each stop is arranged in order, explaining the stories of people, events and objects encountered by the ancestors at each site, and at the same time comprehensively presenting cultural creative products in the exhibition hall. The ten stops are arranged in order: with the

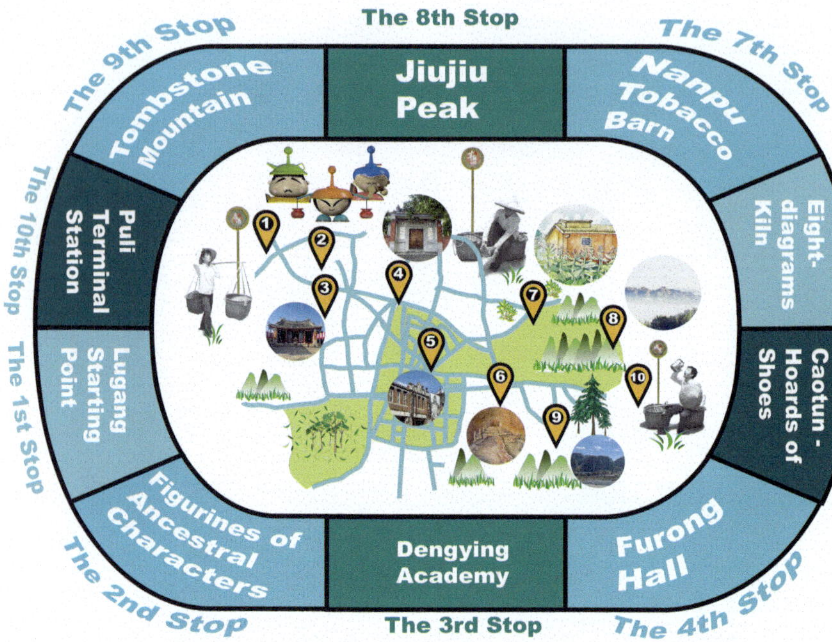

Fig. 5.10 Life-size Caotun Monopoly game display concept

first stop "Lugang" as the starting point, the second stop "traditional dolls," the third stop Dengying Academy, the fourth stop Furong Hall, the fifth stop "Caotun" or mound of straw sandals, the sixth station Eight-diagrams Kiln, the seventh station Nanpu Tobacco Barn, the eighth station JiujiuPeak, the ninth station Tombstone Mountain and the tenth and final station "Puli."

1. The 1st stop: Lugang starting point

"The Household Hardware Selling Young Man Fan Shu Sun (Sweet Potato Son)" Before the sun is even up, he finished the sweet potato breakfast made by his wife. Afterwards, he picks up the tools and the hardware he uses to support his family, puts on his bamboo hat and his straw sandals. Going to and from Lugang takes the man four days. Before embarking on this long journey, he reluctantly says goodbye to his wife and son. Upon arriving at port Lugang and picking up trade goods, he prepares to head towards Puli-she to do business (Figs. 5.15a, b, 5.16 and Table 5.6).

2. The 2nd stop: Figurines of ancestral characters

"The Household Hardware Selling Young Man Fan Shu Sun (Sweet Potato Son)" travels to and from Lukang, Changhua, Kuaiguan, Shekou, Xianzhuang, Beitou, to Cao Xie Dun where he replaces his now worn-out straw sandals. On the way, he will run into the "Straw Shoe Weaving Grandma" in a traditional store and purchase a new pair of straw shoes. After buying a new pair of straw sandals and reaching Puli-she from the east where he can rest. The young man encounters the store manager

Table 5.4 Design and planning of creative products

Site	Place	Product category	Exhibition area content	Scenario content
The 1st stop	Lugang starting point	Exhibition introduction, product series	From the floor to the starting area	Lukang station sign, booth and panels
The 2nd stop	Figurines of ancestral characters	Figurines of Ancestral Characters	From the floor to the doll area	Booth and panels
The 3rd stop	Dengying academy	Candlesticks, bird cages, product series	From the floor to the Dengying academy area	Booth and panels
The 4th stop	Furong hall	Product series	From the floor to the Furong hall area	Booth and panels
The 5th stop	Caotun-Hoards of shoes	Innovative indoor slippers, product series	From the floor to the Caotun shoe mound area	Caotun station sign, tools for making straw sandals, straw, straw sandals, rice, poles, booth and panels
The 6th stop	Eight-diagrams Kiln	Tea set	From the floor to the Bagua Kiln area	Booth and panels
The 7th stop	Nanpu tobacco barn	Ashtray	From the floor to the Nanpu tobacco barn area	Booth and panels
The 8th stop	Jiujiu peak	Doll house, flower vase, product series	From the floor to the Jiujiu peak area	Booth and panels
The 9th stop	Tombstone Mountain	Storage cartridge	From the floor to the Tombstone Mountain area	Booth and panels
The 10th stop	Puli terminal station	Series product, exhibition author introduction, autograph book	From the floor to the terminal station of Puli area	Puli station sign, wine urn, bamboo hat, booth and panels
Middle area		Questionnaires		Booth and panels

"O-ji-sang of Bengfeng Tea Shop" and proceeds to have a great conversation with him, creating one of the scenes in the story. The three characters described above are the heart and soul of this "Lugang Danpushe" story (Figs. 5.17, 5.18 and Table 5.7).

3. The 3rd stop: Dengying academy

"The Household Hardware Selling Young Man Fan Shu Sun (Sweet Potato Son)" heads back to Lugang, via Changhua, Kuaiguan, Shekou, Xianzhuang, and arrived

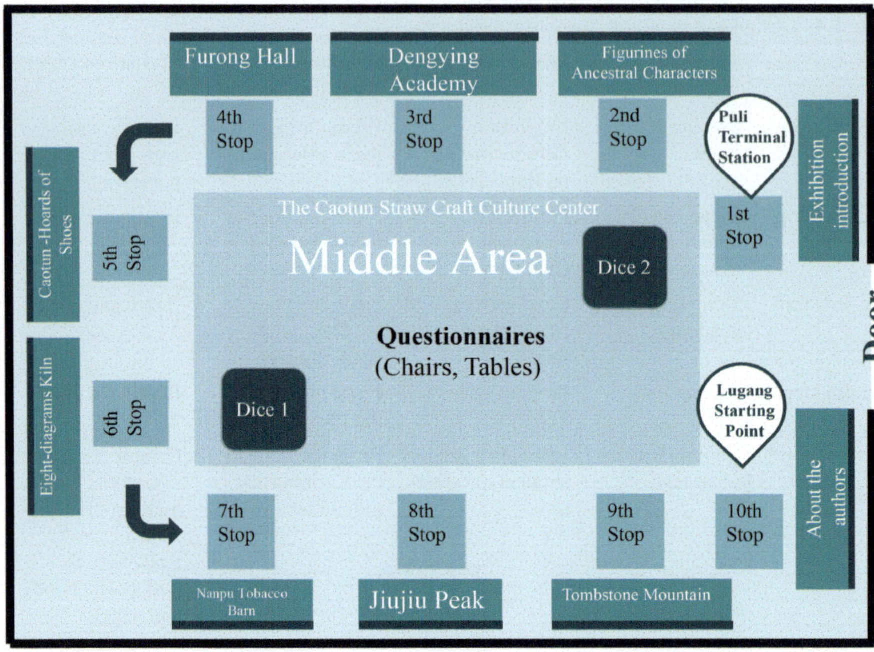

Fig. 5.11 Site layout of the Caotun Straw Craft Culture Center

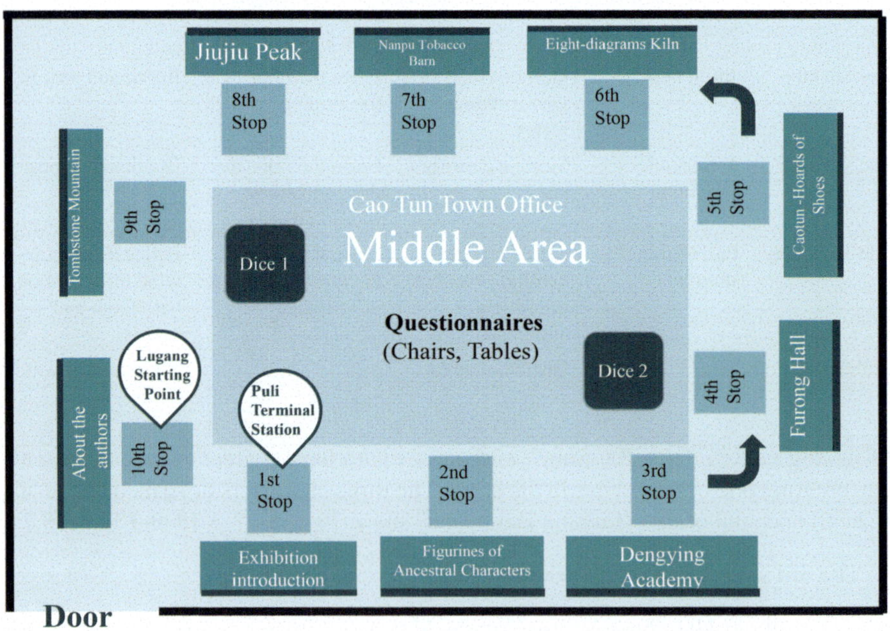

Fig. 5.12 The layout of the exhibition center on the 2F of Caotun Town Office

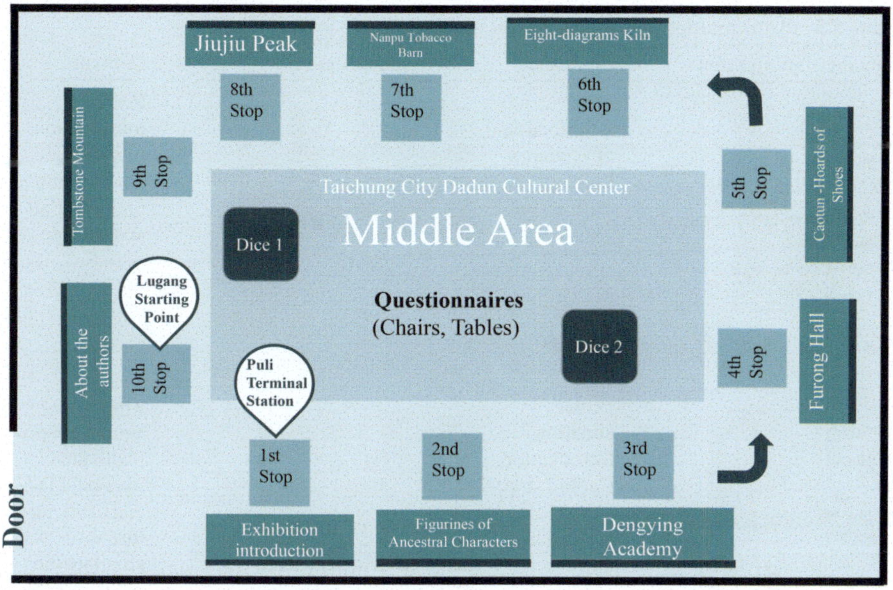

Fig. 5.13 Taichung City Dadun Cultural Center, Wenying building configuration map and site configuration map

(a) **(b)** **(c)**

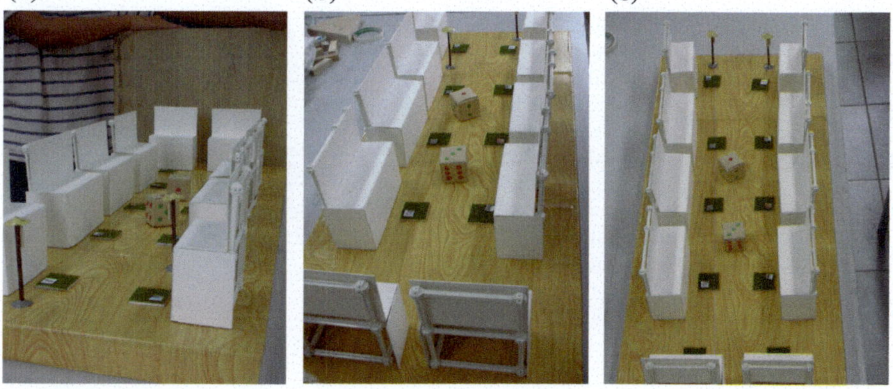

Fig. 5.14 **a–c** Exhibit 1:20 scale model

in Beitou at dusk, passing through the third-level historic site "Dengying Academy," which was a university at that time. In February and August, the Spring and Autumn Festivals are held respectively. Outside the temple, "The Household Hardware Selling Young Man Fan Shu Sun (Sweet Potato Son)" prayed to Emperor Wenchang outside the temple, asking for him to bless his hometown of Lugang and ensure that

Table 5.5 Platforms, decorations, and props used in creative products expos

Lugang starting point	Content	Puli terminal station	Content
	Poster introduction of scenario, Lugang station sign		Poster introduction of scenario, straw, bamboo hat, urn, Puli station sign
Caotun station, where shoes collect into mounds	Content	Site block	Content
	Poster situation introduction, Caotun stop sign, tools for making straw sandals, straw, straw sandals, rice, poles, booth		The square-shaped plank area uses pebbles and artificial turf to simulate the paths that the ancestors walked through at that time
Dice	Content	Booth	Content
	Placed in the middle of the exhibition area, when viewing the exhibition, there is a sense of an interesting presence of large-scale monopoly		Exhibit final cultural creative products

everyone remains safe, before immediately heading out to the next stop in his journey (Figs. 5.19, 5.20, 5.21, 5.22 and Table 5.8).

4. The 4th stop: Furong hall

"The Household Hardware Selling Young Man Fan Shu Sun (Sweet Potato Son)" passes the fourth stop "Furong Hall." In the courtyard, he happened to meet Mr. Hong Lao, the patriarch of Fu Rong, who was busy cultivating plants in the courtyard; "The Household Hardware Selling Young Man Fan Shu Sun (Sweet Potato Son)" praised Mr. Hung and said, "You are a role model for the family of doctors in Caotun.

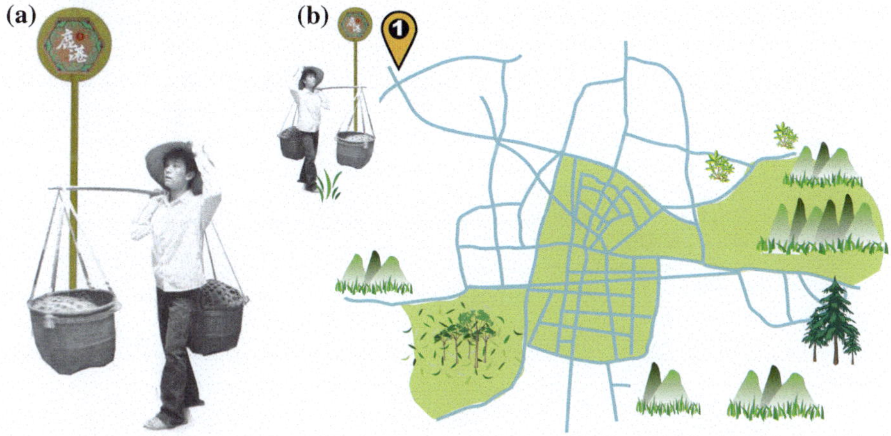

Fig. 5.15 **a**, **b** The 1st stop: Lugang starting point

The children and grandchildren you have cultivated have produced many famous doctors." As the two chat and gossip (Figs. 5.23, 5.24 and Table 5.9).

5. The 5th stop: "Caotun," where shoes have accumulated into mounds

Upon his arrival at the Caotun stop, where shoes have accumulated into mounds, the Household Hardware Selling Young Man Fan Shu Sun buys new straw sandals from the familiar "Straw Sandal Weaving Grandma," and discards his old sandals in Caotun. While doing so the Household Hardware Selling Young Man Fan Shu Sun says to the "Straw Sandal Weaving Grandma," "The straw sandals you weave are really durable, as expected from a master craftswoman!." The straw sandals replaced by the porters accumulated over time in Caotun, forming a "straw sandal mound," which eventually became the place name of "Caotun." It was dark and the "The Household Hardware Selling Young Man Fan Shu Sun" rested in Caotun for the night and waited until dawn to continue his journey towards Puli-she (Figs. 5.25a, b, 5.26, 5.27 and Table 5.10).

6. The 6th stop: Eight-diagrams Kiln

"The Household Hardware Selling Young Man Fan Shu Sun" passes through the sixth station of Eight-diagrams Kiln. At that time, a lot of kiln bricks were produced. In the Eight-diagrams Kiln, there were many brick masters, sweating through their shirts, burning bricks in shifts 24 h a day. The tears of hard work on the master's faces inspired "the Household Hardware Selling Young Man Fan Shu Sun" to also have a positive spirit to fight for a higher social position (Figs. 5.28, 5.29, 5.30 and Table 5.11).

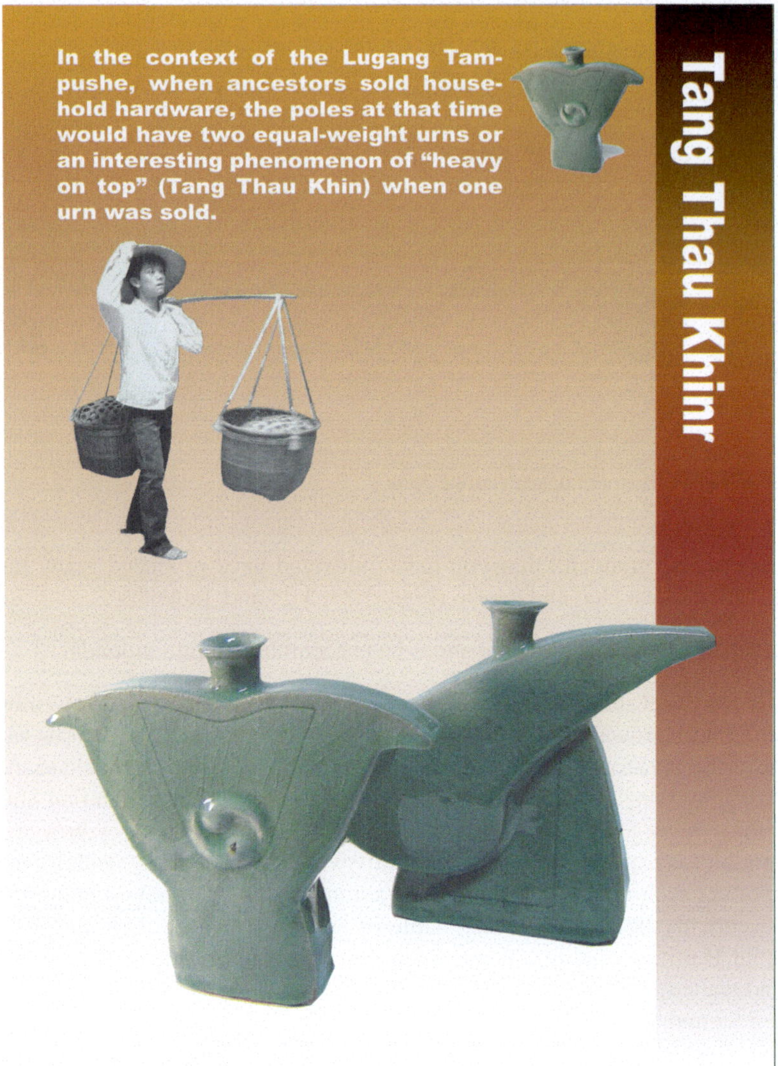

Fig. 5.16 Tang Thau Khin

7. The 7th stop: Nanpu tobacco barn

"The Household Hardware Selling Young Man Fan Shu Sun" passed the seventh stop "Nanpu Tobacco Barn" and saw farmers who were growing tobacco near the Tobacco Barn, harvesting crops under the sun, and cooperating with each other. Farmers did not necessarily get a share of their hard work. In the event of natural disasters, the farmers do not complain about nature and others, continue to cultivate on this land, and encourage the "The Household Hardware Selling Young Man Fan Shu Sun" to

Table 5.6 The 1st stop: Lugang starting point

The 1st stop: Lugang starting point		
	No. 01	Size: 16 cm × 5.6 cm × 18 cm (Upper right)
	Work title: Tang Thau Khin-Cultural Connotation Product Series	
	Design: the first item on the left is an ancestor carrying an even urn. The first item on the right is an item with the design of an ancestry carrying an unbalanced urn, resulting in the Tang Thau Khin (unbalanced) phenomenon	
	Color: green	
	Material: pottery, celadon	
	Function: decorative vase	
	Cultural connotation and meaning: in the context of the Lugang Danpushe, when ancestors sold household hardware, the poles at that time would have two equal-weight urns or an interesting phenomenon of "heavy on top" (Tang Thau Khin) when one urn was sold	

work hard and continue on his journey to the next destination (Figs. 5.31, 5.32 and Table 5.12).

8. The 8th stop: Jiujiu peak

Looking at the "Jiujiu Peaks," one of the eight scenic spots in Taiwan as proclaimed during the Qing Dynasty, "The Household Hardware Selling Young Man Fan Shu Sun" remembered the ancient legend: this peak originally had 100 peaks, causing the ascendance of an emperor, Tiangong (God in the Sky) did not allow this, and struck one peak with lighting, leaving ninety -nine (pronounced Jiujiuin Chinese, an abbreviation of jiu-shi-jiu) peaks left. Thinking of this story, the young man is reminded of his daydream of perhaps becoming an emperor one day. It is also said that Lin Shuangwen, when rebelling against the Qing Dynasty, led troops through this place and buried a lot of gold and silver treasures under the peak. These stories make "the Household Hardware Selling Young Man Fan Shu Sun" day dream about taking these riches and becoming a millionaire in his own right. So the young man warns himself "there's no such thing as free lunch," and stays strong in his belief that earning his share through hard work is more worthwhile as he continues on to his next stop (Figs. 5.33, 5.34, 5.35, 5.36 and Table 5.13).

9. The 9th stop: Tombstone Mountain

"The Household Hardware Selling Young Man Fan Shu Sun" passes through the ninth stop, which is the magnificent "Tombstone Mountain." The geographical location of

Fig. 5.17 The 2nd stop: Figurines of ancestral characters

the mountain is said to protect those living in Lugang and Niumatou (modern day Qingshui). In the twentieth year of the Republic of China, a corner of the mountain collapsed, leaving the region uninhabitable. So local residents bought pigs, sheep, livestock and gold and silver on their behalf, worshipped the gods of Tombstone Mountain and prayed for blessings and peace, and planted bamboo. The community was very peaceful after the incident. The religious "The Household Hardware Selling Young Man Fan Shu Sun," unloads the cargo he is carrying, and clasps his hands together while murmuring prayers at the mountain, praying for the safety and prosperity of the country as well as good weather for the harvests (Figs. 5.37, 5.38 and Table 5.14).

10. The 10th stop: Puli terminal station

After "The Household Hardware Selling Young Man Fan Shu Sun" arrived at the ninth station "Tombstone Mountain," he travelled East to enter "Puli-she," and finally arrived at the terminal station. At the tea shop in front of him, he happened to meet "Ojisan of Bengfeng Tea Shop." The two talked about: "'Rome was not built in a day,' and the same is true for 'Lugang Danpushe.' We work day and night, every day, step-by-step, working hard to support our family, working day and night, sweating through our clothes, show a down-to-earth spirit, and it is worth learning from for our descendants." After drinking tea, "The Household Hardware Selling Young Man

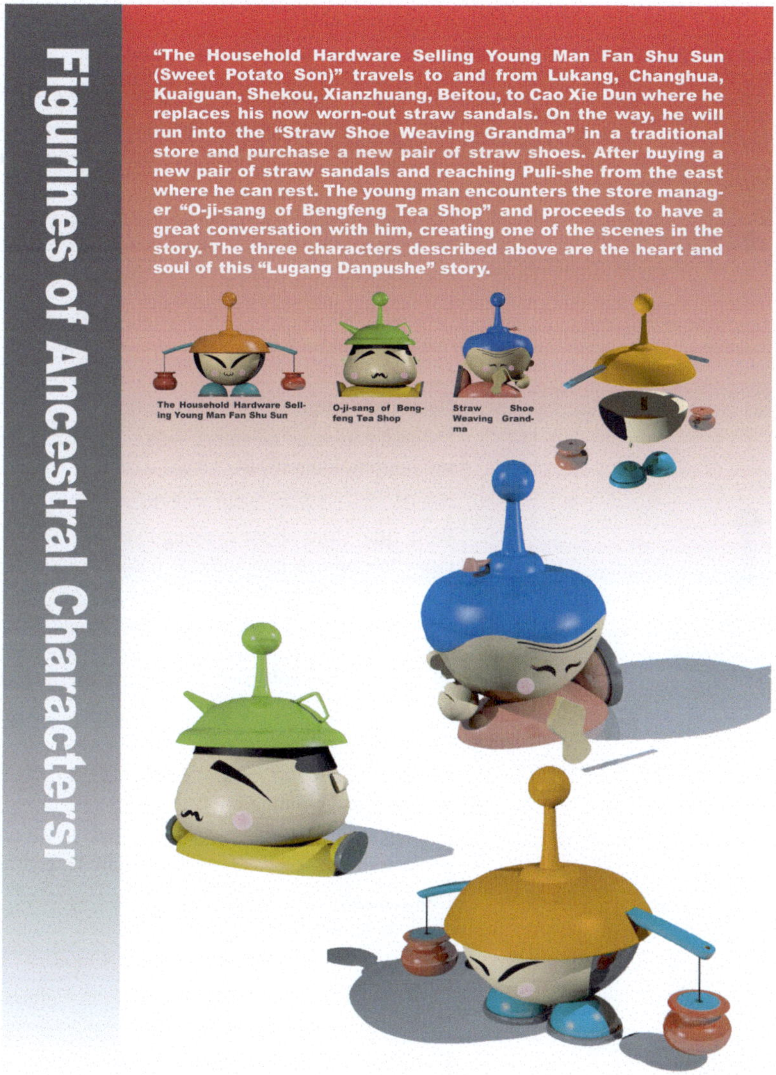

Fig. 5.18 Doll of Caotun ancestors

Fan Shu Sun" also sold out the household hardware on the way, and delivered the goods to Puli-she, and then purchased the local products and camphor and other goods for the return journey. The round trip back to Lugang will take another four days (Figs. 5.39a, b, 5.40 and Table 5.15).

In this book, the main story of "Lugang Danpushe" is planned by using the scenario-based design method, the static exhibition is added to the rich story, and thus

Table 5.7 The 2nd stop: Figurines of ancestral characters

The 2nd stop: Figurines of ancestral characters		
	No. 02	Size: 16 cm × 8 cm × 13 cm
	Work title: The Household Hardware Selling Young Man Fan Shu Sun—cultural concreteness product series	
	Form: against the background of the Lukang Danpushe, a porter with a hat, a pole and two urns	
	Color: orange, dark green, skin tone and red	
	Material: using CNC model making, ABS raw materials and painting	
	Function: the head, face, feet, urn and pole of the doll can be combined by DIY	
	Cultural extension meaning: dolls of ancestors who sell household hardware	
	No. 03	Size: 9 cm × 5 cm × 13 cm
	Work title: straw shoes weaving Fan Shu Grandma-cultural concreteness product series	
	Form: in the context of the era of the Lugang Danpushe, women took up the work of making straw sandals, making it a cottage industry. This product takes the way people used to weave straw sandals into dolls	
	Color: blue, skin tone, pink	
	Material: using CNC model making, ABS raw materials and painting	
	Function: the head, face, hairpin, hands, straw shoes and feet of the doll can be combined by DIY	
	Cultural extension meaning: Dolls of Ancestors weaving straw shoes	
	No. 04	Size: 9 cm × 5 cm × 13 cm
	Work title: Ojisan at Bengfeng Tea Shop-Cultural Concreteness Product Series	
	Form: in the context of the era of the Lugang Danpushe, it describes the tea shop where the ancestors rested and drank tea, showing amusing teapot hats and friendly local residents	
	Color: turquoise, skin tone and yellow	
	Material: using CNC model making, ABS raw materials and painting	
	Function: the teapot hat, face, feet and other parts of the doll can be combined by DIY	
	Cultural extension meaning: ancestor Ojisan who is resting and making tea at the tea shop	

Fig. 5.19 The 3rd stop: Dengying academy

highlighting the three cultural creative product exhibitions. The following section will conduct quantitative analysis on two groups of test subjects: "foreign tourists" and "local residents," for their understanding and evaluations of the cultural creative product exhibition.

5.5 Cognitive Appraisal and Analysis of Cultural and Creative Product Exhibitions

The following section will analyze the results of using academic literature relevant to the scenario-oriented design method to plan out three cultural creative product exhibitions; through using quantitative questionnaires to understand the degree of local cultural understanding as well as understanding of the cultural creative product exhibition and the overall evaluation towards products of both foreign tourists and local residents.

Fig. 5.20 Dengying book case

5.5.1 Questionnaire

1. Questionnaire title: Evaluation and investigation of culture creative commodity exhibitions on local cultural understanding.
2. Tested identities: Surveys are conducted on two different identities: "foreign tourists" and "local residents."

Fig. 5.21 Pious heart

Fig. 5.22 Valedictorian candle tray

3. Test locations: The three test sites are "Taichung City Dadun Cultural Center, Wenying Building," "Caotun Straw Craft Culture Center" and "Caotun Township Public Library," all of which are static exhibition centers. Among them, the

Table 5.8 The 3rd stop: Dengying academy

The 3rd stop: Dengying academy		
	No. 05	Size: 45 cm × 30 cm × 60 cm
	Work title: Dengying book case-cultural concreteness product series	
	Form: the shape is mainly designed with the appearance of the Dengying Academy building	
	Color: the color of the bird cage imitates the color system of the original materials of wood, wood carving and clay sculpture in Dengying academy	
	Material: Dengying academy is an ancient academy, and the building uses cypress, cochin pottery and clipping, etc., so this product uses materials such as pottery, wooden strips, bamboo, rattan and wooden boards	
	Function: in February and August every year, the spring and autumn festivals are held in each respective month. In the festival, the childbirth basket of the four ancient rituals is converted into bird feed, the deacon card is converted into a birdcage door, the Kuixing magic pen is converted into a bamboo pole on which the bird stands, and the writing paper is converted to a water dispenser	
	Cultural extension meaning: in the spring and autumn festivals of Dengying academy, the four ancient ritual utensils were used to transform the scent of books in the cage and the concept of free flying in the cage	
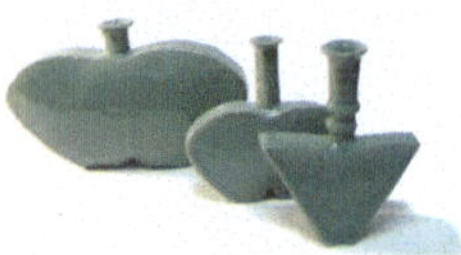	No. 06	Size: 15 cm × 5.3 cm × 25 cm(Upper left)
	Work title: Pious heart-cultural connotation product series	
	Form: to express the devout hearts of believers, arrange heart-shaped designs of various sizes	
	Color: green	
	Material: pottery, celadon	
	Function: decorative vase	
	Cultural connotation and meaning: Dengying academy holds two spring and autumn seasons in February and August every year. In the festival, it is a prayer for descendants to make improvements and rank high in their academic classes	
	No. 07	Size: 7.5 cm × 6 cm × 7 cm
	Work title: Valedictorian candle tray-cultural concreteness product series	
	Form: the chinchilla symbol in Dengying academy is applied to the surface of the candle tray	
	Color: green	
	Material: pottery, celadon	
	Function: convert the bright lamps that believers lit in Dengying academy into candle lamps	

(continued)

Table 5.8 (continued)

The 3rd stop: Dengying academy	
	Cultural extension meaning: the concept is taken from the local residents seeking bright futures for their children

Fig. 5.23 The 4th stop: Furong hall

exhibition venue of "Caotun Township Straw Craft Culture Museum" has a sales area for cultural creative products as well as a DIY creation area (Table 5.16).

4. Questionnaire purpose: To verify the effectiveness of the script-guided exhibition, to understand the close evaluation of cultural creative products by foreign tourists and local residents, and to analyze the survey of the exhibition's awareness of local culture.
5. Open-ended questionnaire: Because it is to investigate the relationship between foreign tourists and local residents, the identity of the respondents is not limited, and open-ended questionnaires are conducted. Anyone who visits this exhibition is eligible to receive the questionnaire.

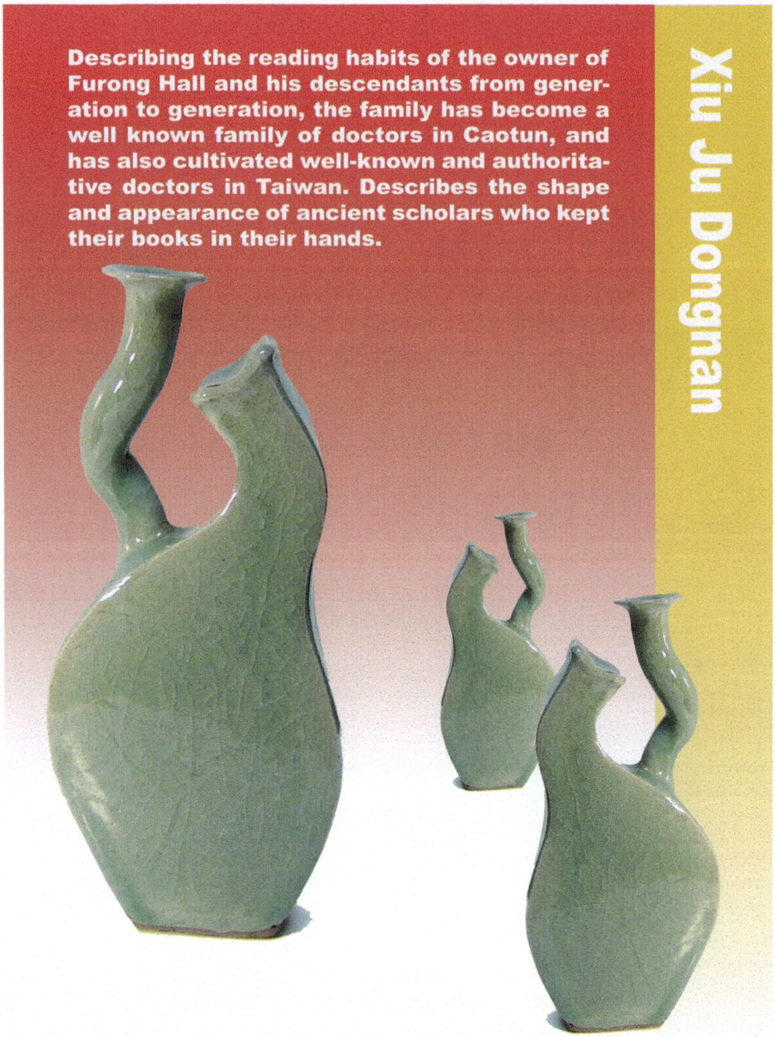

Fig. 5.24 Xiu Ju Dongnan

6. Analysis focus: Cognition analysis of the impression of Caotun cultural resources, analysis of cultural cognition and evaluation of culture creative commodity exhibitions, analysis of the effectiveness of the scenario-based design method, analysis of the definition and status of local cultural creative products within the minds of consumers, among other things.

Table 5.9 The 4th stop: Furong hall

The 4th stop: Furong hall		
	No. 08	Size: 13 cm × 6 cm × 16 cm
	Work title: Xiu Ju Dongnan-cultural concreteness product series	
	Form: describes the shape and appearance of ancient scholars who kept their books in their hands	
	Color: green	
	Material: pottery, celadon	
	Function: decorative vase	
	Cultural connotation and meaning: describing the reading habits of the owner of Furong hall and his descendants from generation to generation, the family has become a well known family of doctors in Caotun, and has also cultivated well-known and authoritative doctors in Taiwan	

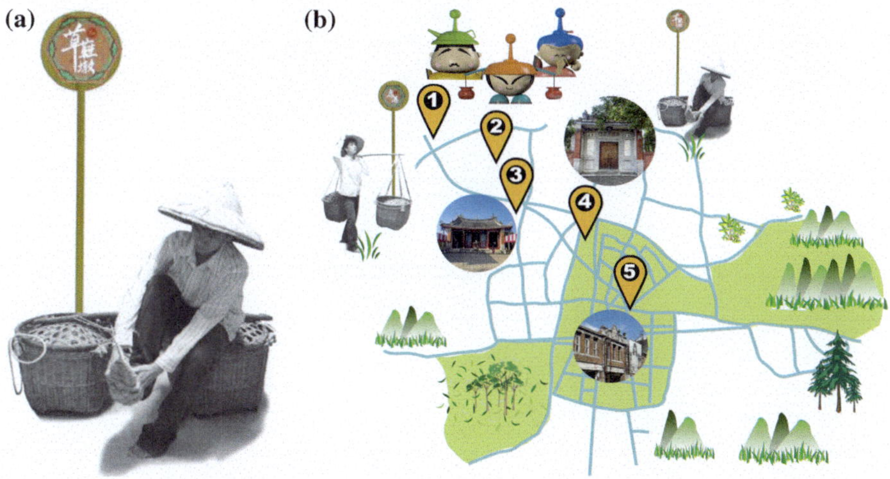

Fig. 5.25 **a**, **b** 5th stop: Hoards of shoes

5.5.2 *Basic Data Analysis of Subjects*

A total of 370 valid questionnaires were conducted in this survey, which were conducted in three areas, namely, "Taichung City Dadun Cultural Center, Wenying Building," "Caotun Straw Craft Culture Center" and "Caotun Township Public

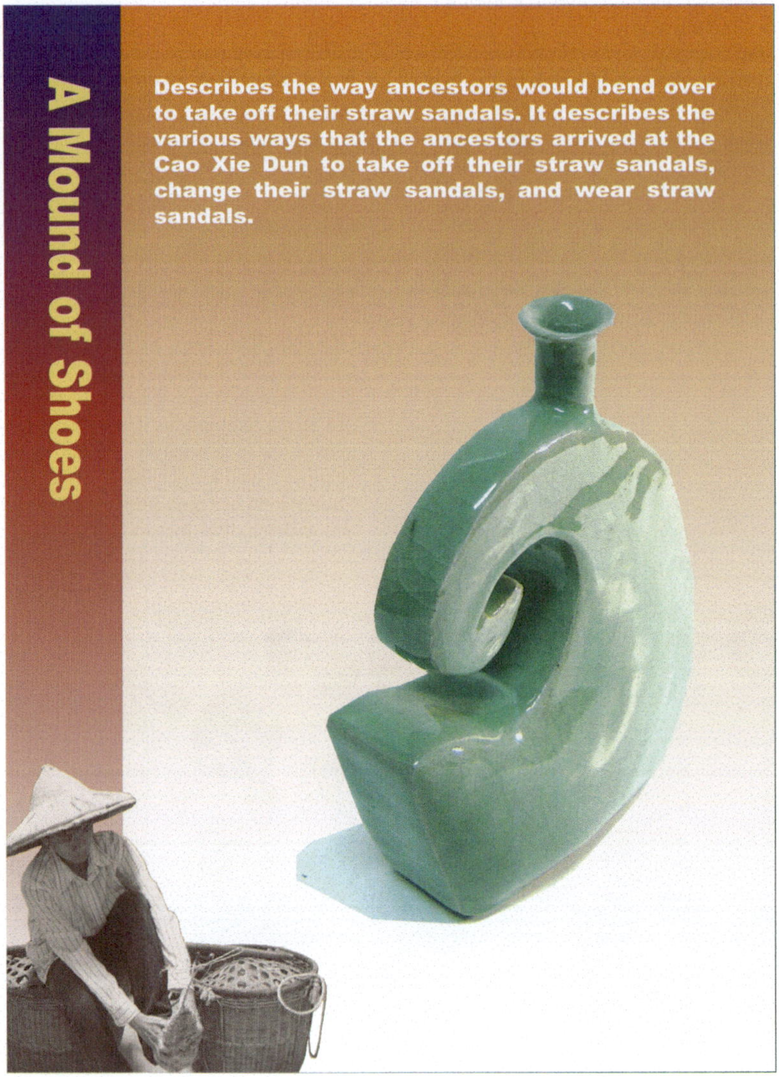

Fig. 5.26 A mound of shoes

Library," and 370 valid questionnaires were collected. There were 127 "foreign tourists" and 243 "local residents," making a total of 370 subjects. Among them, 148 were "males" (40%) and 222 were "females" (60%). 190 people (51.3%) participated in the exhibition with an education level below high school or vocational school level, and 180 people (48.7%) were above junior college. In terms of "employment status" 152 (41.1%) were students, accounting for more than half of total respondents; 93

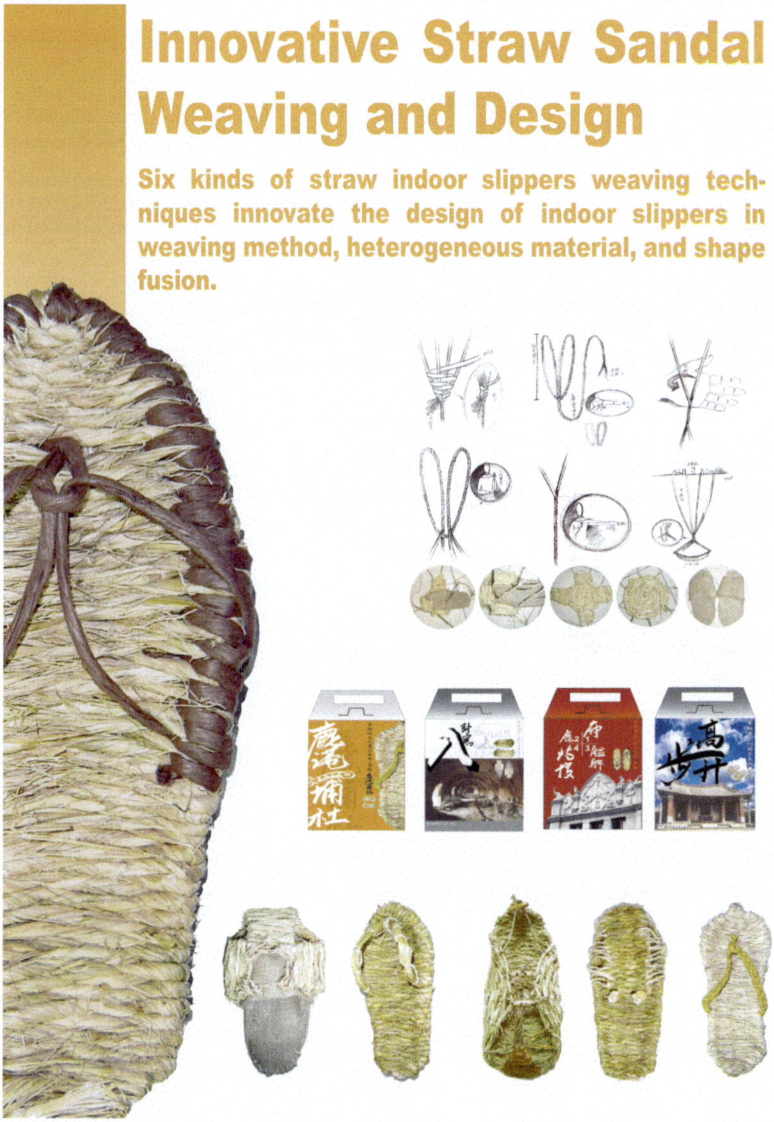

Fig. 5.27 Innovative straw sandal weaving and design

(25.1%) were in the military, public service, or a teacher, 20 (5.4%) in freelance work, 30 in cultural organizations (8.1%), 24 in scholars, farmers, businessmen (6.5%), and 51 in the service sector (13.8%) (Table 5.17).

Table 5.10 The 5th stop: Hoards of shoes

The 5th stop: Hoards of Shoes		
	No. 09	Size: 13 cm × 6 cm × 21 cm
	Work title: a mound of shoes-cultural concreteness product series	
	Form: describes the way ancestors would bend over to take off their straw sandals	
	Color: green	
	Function: decorative vase	
	Material: pottery celadon	
	Cultural connotation and meaning: it describes the various ways that the ancestors arrived at the Cao Xie Dun to take off their straw sandals, change their straw sandals, and wear straw sandals	
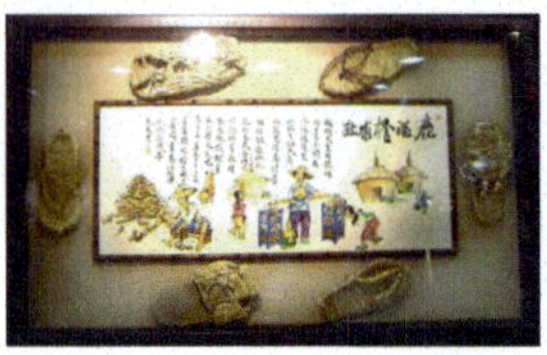	No. 10	Size: 180 cm × 10 cm × 92 cm (Outer frame)
	Work title: indoor healthy straw slippers-cultural concreteness product series	
	Form: slippers, flip flops and half-covered sandals, etc.	
	Color: original color of straw material	
	Material: straw, paper rattan, cloth and hemp rope	
	Function: indoor slippers	
	Cultural extension meaning: six kinds of straw indoor slippers weaving techniques innovate the design of indoor slippers in weaving method, heterogeneous material, and shape fusion	

5.5.3 Analysis of Cognition of Caotun Cultural Resource Impression

1. What do you think is the most representative spiritual symbol of Caotun?

The statistical method used in this section is taking the mean. The maximum score a respondent can give is a 5 out of 5, with 1 representing strongly disagree, 2 representing disagree, 3 representing acceptable, 4 representing agree, and 5 representing strongly agree. It shows that foreign tourists have a high impression of the "National Taiwan Craft Research and Development Institute" with an average of 4.3 points, indicating that the National Taiwan Craft Research and Development Institute is the most impressive place in the hearts of foreign tourists due to frequent nationwide activities. After the Institute, in second is the "JiujiuPeaks" with an average score of 4.16, which is the most familiar to foreign tourists. The local residents have a high average score of 4.9 for "straw sandals," which belongs to the traditional crafts

Fig. 5.28 The 6th stop: Eight-diagrams Kiln

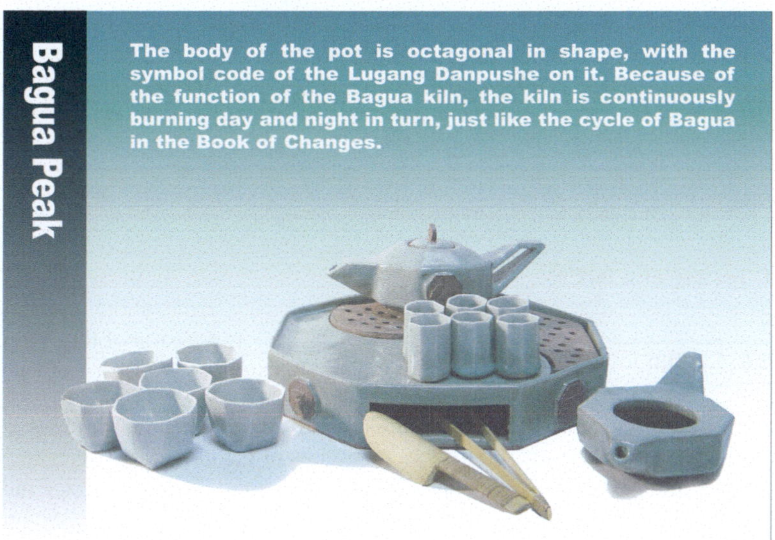

Fig. 5.29 Bagua Peak

Fig. 5.30 Tea fragrance Tai Chi

of Caotun. Caotun town records record the history of "straw sandal" weaving, and also weaving the place name of Caotun. The data shows that the respondents have a high recognition of the spiritual symbol of Caotun in straw sandals. Following straw sandals themselves, the "straw" industrial resources of Caotun had an average score of 4.12 points, indicating that the local residents understand the agrarian rice industry in Caotun. In addition, in terms of the statistic regarding Caotun's spiritual symbol, "foreign tourists" and "local residents" gave "straw sandals" the highest average in 4.35 (Fig. 5.41 and Table 5.18).

Table 5.11 The 6th stop: Eight-diagrams Kiln

<table>
<tr><th colspan="3">The 6th stop: Eight-diagrams Kiln</th></tr>
<tr><td rowspan="7"></td><td>No. 11</td><td>Size: 65 cm × 30 cm × 15 cm (A set)</td></tr>
<tr><td colspan="2">Work title: Bagua Peak-cultural concreteness product series</td></tr>
<tr><td colspan="2">Form: combinations such as teapots, cups and tea trays are all designed in the shape of the eight trigrams</td></tr>
<tr><td colspan="2">Color: green</td></tr>
<tr><td colspan="2">Material: pottery, celadon, bamboo, rattan</td></tr>
<tr><td colspan="2">Function: tea set</td></tr>
<tr><td colspan="2">Cultural extension meaning: the body of the pot is octagonal in shape, with the symbol code of the Lugang Danpushe on it. Because of the function of the Eight-diagrams Kiln, the kiln is continuously burning day and night in turn, just like the cycle of Bagua in the Book of Changes</td></tr>
<tr><td rowspan="7"></td><td>No. 12</td><td>Size: 56 cm × 50 cm × 20 cm (A set)</td></tr>
<tr><td colspan="2">Work title: tea fragrance Tai Chi-cultural concreteness product series</td></tr>
<tr><td colspan="2">Form: Tai Chi has a harmonious form of yin and yang</td></tr>
<tr><td colspan="2">Color: green</td></tr>
<tr><td colspan="2">Material: pottery, celadon, bamboo, rattan</td></tr>
<tr><td colspan="2">Function: tea set</td></tr>
<tr><td colspan="2">Cultural extension meaning: Bagua means the harmony of yin and yang of Tai Chi, which is like a good tea brewed in the tea ceremony, which can harmonize the aroma</td></tr>
</table>

2. What is the origin of the place name of Caotun?

There are different theories about the name of Caotun, according to the records of *Caotun Township Cronicle*: "During the Qianlong and Jiaqing years in the late Qing Dynasty, the predecessors of the Lugang Danpushe went to and from Puli. These settlers, porters, and business travelers would arrive in Cao Xie Dun, the land where they would exchange or discard their old straw sandals; over time, the abandoned straw sandals form mounds." "Straw sandals" has been weaved into the place name of Caotun. As for the opinions regarding the place name in the hearts of the "foreign tourists" and "local residents," those who choose "Lugang Danpushe, the ancestors went to and from Lugang and Puli, these settlers, porters, and business travelers would arrive in Cao Xie Dun, the land where they would exchange or discard their old straw sandals," the average is 4.37. The average value is 3.7 for "abandoning old straw sandals for new ones, and the straw sandals become mounds over time." The average value of "the topographical map of Caotun is shaped similar to straw sandals and thus named as straw sandals" is 2.62 and ranked third; a minority who

Fig. 5.31 The 7th stop: Nanpu tobacco barn

chose "I don't know about the origin of the name 'Caotun'" also had an average of 1.21. The data proves that "foreign tourists" and "local residents" have considerable knowledge of the origin of the basic Caotun place names, which confirms that the scenario-oriented design method for planning the exhibition which takes "Lugang Danpushe" as the background story, after quantitative analysis, to have been able to bring a high degree of familiarity with this story, deepening the impression of the cultural creative product exhibition (Fig. 5.42 and Table 5.19).

3. Based on your understanding of Caotun's natural, production, human, cultural and landscape resources, which one is the most representative of the township itself?

Statistics show that the average value of Caotun's "production resources" is 4.16, making it the public's favorite choice. Production resources contain items such as: rice, straw, straw sandals, Pinglin grapes, Pinglin lychees, Shuangdong betel nuts, and others, in Caotun. The average value of "landscape resources" is 3.59, which is good for second place among the five resources, landscape resources include items such as: JiujiuPeaks in Caotun, Eight-diagrams Kiln, Time-honored Camphor Tree, Dengying Academy, among others. After the knowledge base is established for this

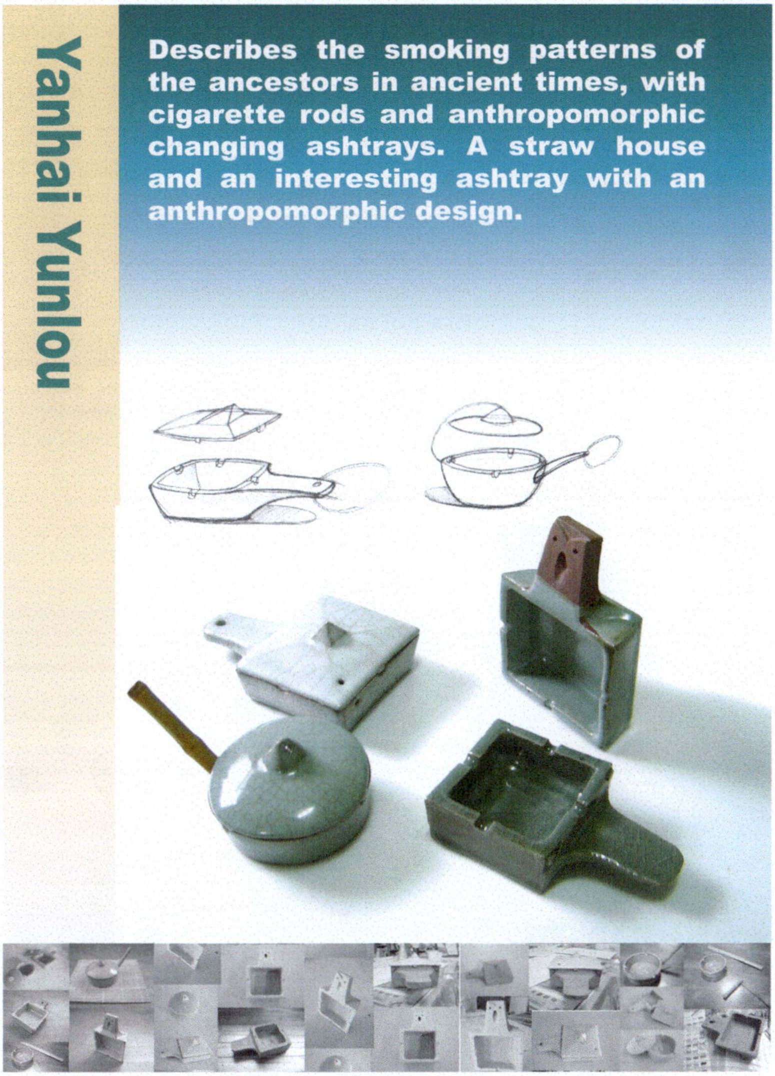

Fig. 5.32 Yanhai Yunlou

statistical purpose, the direction of inputting data in the future can refer to selecting production resources and landscape resources as the primary goals (Fig. 5.43 and Table 5.20).

4. What is your impression of the landscape resources in Caotun?

For the tourism industry, the local landscape and scenery are the biggest incentive for attracting people. According to the statistics of the respondents' impression of Caotun's scenic resources, the average value of the option "Jiujiu Peaks" is 4.11,

Table 5.12 The 7th stop: Nanpu tobacco barn

The 7th stop: Nanpu tobacco barn		
	No. 13	Size: 13 cm × 6 cm × 16 cm (Upper right)
	Work title: Yanhai Yunlou-Cultural Concreteness Product Series	
	Form: a straw house and an interesting ashtray with an anthropomorphic design	
	Color: green	
	Material: pottery, celadon, bamboo	
	Function: ashtray	
	Cultural extension meaning: describes the smoking patterns of the ancestors in ancient times, with cigarette rods and anthropomorphic changing ashtrays	

Fig. 5.33 The 8th stop: Jiujiu peak

which shows that the general public has a high opinion of Caotun's "Jiujiu Peaks." The impression of "Dengying Academy" ranks second with an average of 3.76, and the impression of "Eight-diagrams Kiln" with an average of 3.19 ranks third. The most unimpressive option in the scenic part is the average value of "Furong Hall"

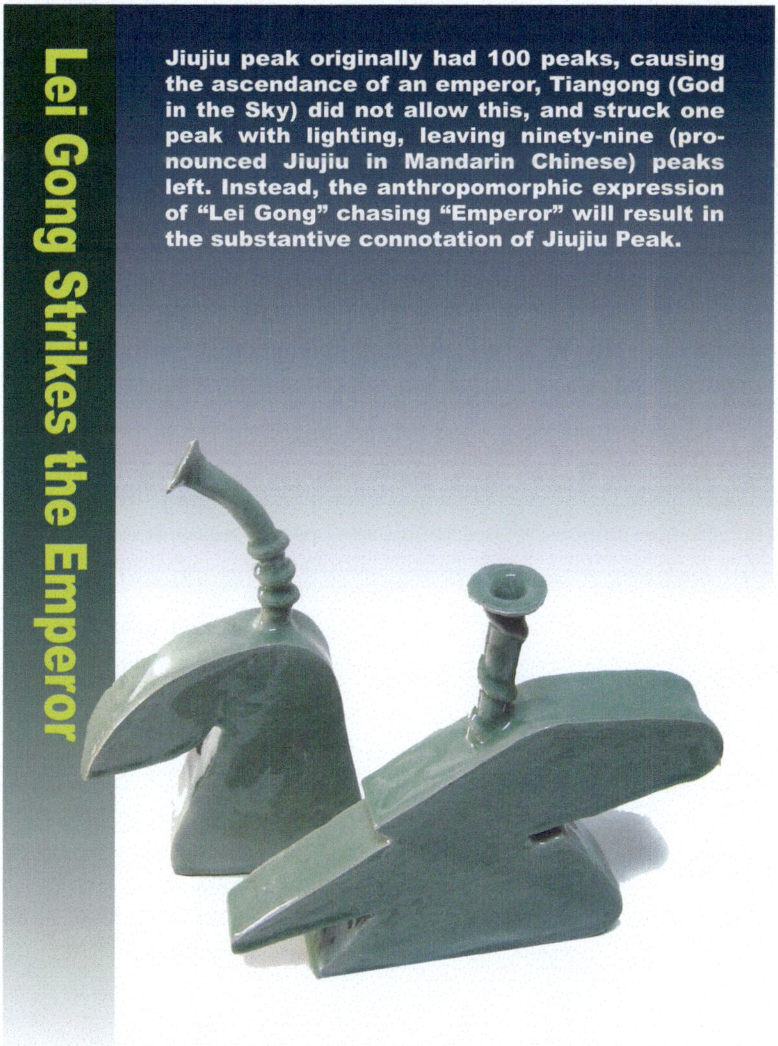

Fig. 5.34 Lei Gong strikes the emperor

2.79. This place has been renovated and maintained by the government and local cultural groups as a new scenic spot in Caotun, and is the most unfamiliar to the subjects (Fig. 5.44 and Table 5.21).

Fig. 5.35 Cao Xie Dun tea house

5.5.4 Consumption and Analysis of the Cultural Creative Products on Exhibit

1. Are you willing to purchase existing local cultural creative products in Caotun Township?

Fig. 5.36 Jiujiu peak

The average value of the option "Very willing" was 2.07, and the average value of "Very unwilling" was 4.95. This data confirms that "very unwilling" has more than double the average value of "very willing," indicating that most of the respondents are not very willing to purchase the existing Caotun cultural creative products.

2. If the cultural creative products displayed at this exhibition enter the broader market in the future, will you choose to buy them?

Table 5.13 The 8th stop: Jiujiu peak

The 8th stop: Jiujiu peak		
	No. 14	Size: 23 cm × 12 cm × 57 cm (Lower right)
	Work title: Lei Gong strikes the emperor-cultural connotation product series	
	Form: the one on the right portrays the thunder pattern as "Lei Gong," and the one on the left is the "Emperor," who is being chased, with his head down for fear of being struck by thunder	
	Color: green	
	Material: pottery, celadon	
	Function: decorative vase	
	Cultural connotation and meaning: Jiujiupeak originally had 100 peaks, causing the ascendance of an emperor, Tiangong (God in the Sky) did not allow this, and struck one peak with lighting, leaving ninety-nine (pronounced Jiujiuin Mandarin Chinese) peaks left. Instead, the anthropomorphic expression of "Lei Gong" chasing "Emperor" will result in the substantive connotation of JiujiuPeak	
	No. 15	Size: 40 cm × 50 cm × 60 cm
	Work title: Cao Xie Dun tea house (1:20 scale model)-cultural concreteness product series	
	Form: the nine-nine peaks are regarded as a ridge horizontally and the main body of the peaks is formed on the side	
	Color: the color of the straw itself	
	Material: most of them are mainly made of straw, hemp cloth, artificial turf, all kinds of cardboard, plastic hemp rope, and acrylic	
	Function: decorative doll house	
	Cultural extension meaning: Caotun started with agriculture, and the straw house has many memories for many people. The work combines the style of a straw house and Jiu-jiu-feng into a tea house	

(continued)

Table 5.13 (continued)

The 8th stop: Jiujiu peak		
	No. 16	Size: 47 cm × 32 cm × 32 cm
	Work title: Jiujiu peak-cultural concreteness product series	
	Form: when viewed horizontally as a ridge and a peak on the side, it has a different artistic conception from far and near	
	Color: after the 921 Earthquake in Taiwan, there is a natural yellow color	
	Material: straw, rattan, pottery	
	Function: it is listed as a natural and cultural asset protection area by the state, and it is converted into a flower vessel, which has the function of holding/collecting water; the product is woven with straw and rattan on the ground, which can increase the texture level	
	Cultural extension meaning: after the 921 Earthquake, the Jiujiu peak of the Huangtu mountains was listed as a nature reserve. It is seen as a peak on the side of the ridge, with different shapes appearing as one observes it from different angles far and near	

This option examines whether or not cultural creative products designed using the cultural and historical design symbol knowledge base will, when displayed through the scenario-based design method, will increase the desire to purchase these products among the test subjects. From the options and the results, we can see that the average for "very willing" to purchase is 4.25, and the average for "very unwilling" is 1.26. These averages prove that a majority of test subjects will choose to purchase products displayed within the exhibition (Table 5.22).

5.5.5 *Analysis of the Cognition Towards the Cultural Creative Products Exhibition*

1. How much did you enjoy the method of viewing exhibits, walking maps, telling stories, and understanding culture as a way of presenting the Caotun cultural creative products exhibition?

Fig. 5.37 The 9th stop: Tombstone Mountain

(1) Factor 1: It belongs to foreign tourists and local residents, who directly prefer to choose culture creative commodity exhibitions. Therefore, the factors such as option panel design, product design, and overall planning are named "static display." Statistical results show that the respondents have a very high evaluation of factor one, which is the average of 4.59 for product design, 4.57 for overall planning, and 4.46 for panel design.

(2) Factor 2: The line of movement planning and guide commentary in the options are named "Dynamic Commentary," which refers to the evaluation of the subjects' preference for culture creative exhibitions, cultural communication, and interaction of lines of movement. The data shows that the average value of the moving line planning is 4.51, and the average value of the guide commentary is 4.49; the dynamic walking map is used to understand Caotun's culture, and the story guide is more profound, which verifies the display effect of viewing exhibitions, walking maps, telling stories and understanding culture.

(3) Factor 3: The option "Story Application," the factor is named "Scenario Guidance," the average value is 4.62, which is the highest evaluation among the average values. It is affected by the planning of the scenario-based design method, and the effect of the stories and situations expressed in the display.

(4) Among the above three factor analysis groups, the average value of scenario guidance is 81.950, the average value of dynamic commentary is 60.107,

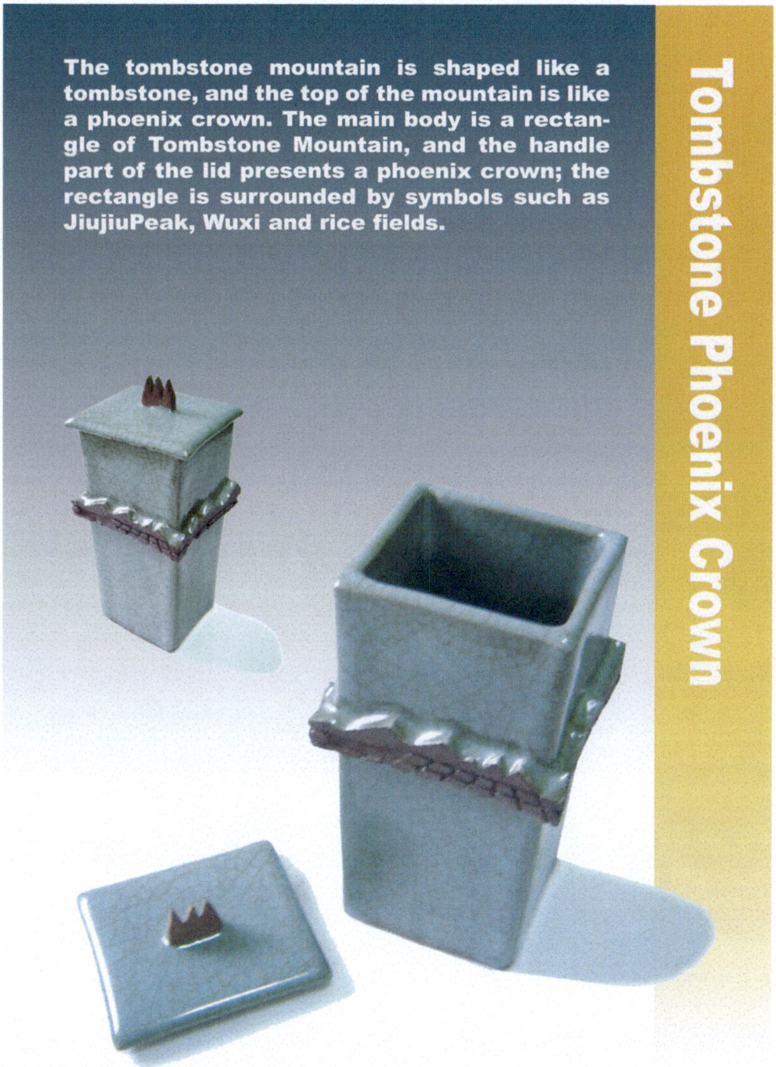

Fig. 5.38 Tombstone phoenix crown

and the average value of static display is 33.591. Effectively achieving the curatorial effect of viewing exhibitions, walking maps, telling stories and understanding culture (Table 5.23).

2. Your evaluation of the adjective vocabulary of this cultural creative product display

   (1) Factor 1: The adjective vocabulary regarding value, aesthetics, grace, and modernity were grouped together and named as a "stylistic preference." In

Table 5.14 The 9th stop: tombstone Mountain

The 9th stop: Tombstone Mountain		
	No. 17	Size: 32 cm × 20 cm × 12 cm
	Work title: Tombstone Phoenix crown-cultural concreteness product series	
	Form: the main body is a rectangle of Tombstone Mountain, and the handle part of the lid presents a phoenix crown; the rectangle is surrounded by symbols such as JiujiuPeak, Wuxi and rice fields	
	Color: green	
	Material: pottery, celadon	
	Function: multifunctional storage cartridge	
	Cultural extension meaning: the Tombstone Mountain is shaped like a tombstone, and the top of the mountain is like a phoenix crown	

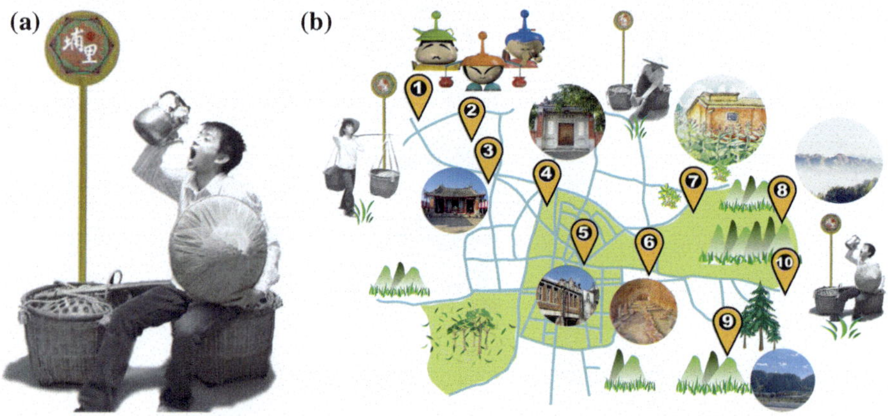

Fig. 5.39 **a**, **b** The 10th stop: Puli terminal station

terms of the first group, "stylistic preference," aesthetics and value, with averages of 3.97 and 3.95 respectively, ranked highest. Indicating that products are tangible, while service is intangible, and experiences are valuable (Pine II & Gilmore, 2013). The value of achieving local cultural recognition through storytelling with cultural creative products is the goal of the scenario-based design of the cultural creative products exhibition. The average for vocabulary regarding modernity had the lowest average of 3.62, with the reason being this option has a conflict with the display of local products.

(2) Factor 2: Subjects' understanding of local culture in the culture creative commodity exhibition, and naming connotative, classic, unique, symbolic and other adjective vocabulary as "cultural understanding." The average

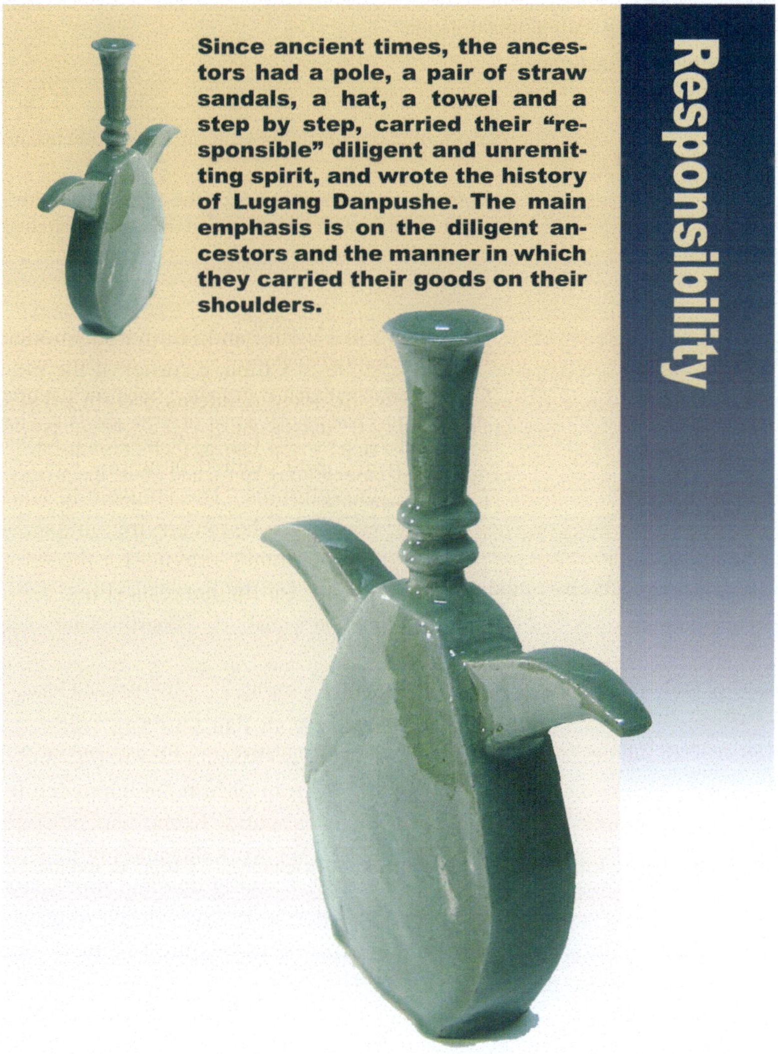

Fig. 5.40 Responsibility

value of connotation related adjectives of the second group of "cultural understanding" is 4.00, the local cultural background is strong, and the connotation narrative of each station story is highly influential.

(3) Factor 3: The respondents' perception of the local traditional culture in the culture creative commodity exhibition, and named the simple, local, and practical adjectives as "traditional imagery." The third group of "traditional imagery" shows that the average value of words related to simple (or simplicity) is 4.00, the average value of product reality is 3.46, and the

Table 5.15 The 10th stop: Puli terminal station

The 10th stop: Puli terminal station		
	No. 18	Size: 18 cm × 25 cm × 32 cm
	Work title: responsibility-cultural connotation product series	
	Form: the main emphasis is on the diligent ancestors and the manner in which they carried their goods on their shoulders	
	Color: green	
	Material: pottery, celadon	
	Function: decorative vase	
	Cultural connotation and meaning: since ancient times, the ancestors had a pole, a pair of straw sandals, a hat, a towel and a step by step, carried their "responsible" diligent and unremitting spirit, and wrote the history of Lugang Danpushe	

Table 5.16 Locations of creative products expos

Questionnaire location	Exhibition venue and spatial properties	Questionnaire subject
Taichung City Dadun Cultural Center, Wenying Building	Purely for exhibition viewing	Tourists and locals
Caotun Straw Craft Culture Center	Sales area for cultural creative products DIY area for cultural creative products	Tourists and locals
Caotun Township Public Library	Purely for exhibition viewing	Tourists and locals

low evaluation shows that the functions of the cultural creative products displayed tend to be decorative (Table 5.24).

5.5.6 Effect Analysis of Scenario-Based Method

1. Do you think your cultural awareness of Caotun has improved after experiencing this cultural creative products exhibition?

The data proves that the exhibition has improved cultural awareness. The average of "strongly agree" is 4.88, the average of "agree" is 4.12; the average of "normal" is 2.68; the average of "disagree" is 1.26, and the average of "strongly disagree" The value is 1.07. The exhibition emphasizes that it is related to the goals of "viewing the

Table 5.17 Cross tabulation analysis: respondents' basic information

Option		Identity					
		Tourists		Locals		Total	
		Number of people	Proportion (%)	Number of people	Proportion (%)	Number of people	Proportion (%)
Gender	Male	56	15.1	92	24.9	148	40
	Female	71	19.2	151	40.8	222	60
Age	Under 29	74	20	157	42.4	231	62.4
	Over 30 years old	53	14.3	86	23.3	139	37.6
An education level	Below high school or vocational school	29	7.8	161	43.5	190	51.3
	College or above	98	26.5	82	22.2	180	48.7
In terms of current employment	Students	61	16.5	91	24.6	152	41.1
	Military, public service, or a teacher	36	9.7	57	15.4	93	25.1
	Freelance work	0	0	20	5.4	20	5.4
	Cultural organizations	17	4.6	13	3.5	30	8.1
	Scholars, farmers, businessmen	0	0	24	6.5	24	6.5
	Service sector	13	3.5	38	10.3	51	13.8
Total		127	34.3	243	65.7	370	100

exhibition, walking the map, telling stories, and understanding the culture," which encourages the subjects to have a deeper understanding and cognition of Caotun's culture (Fig. 5.45 and Table 5.25).

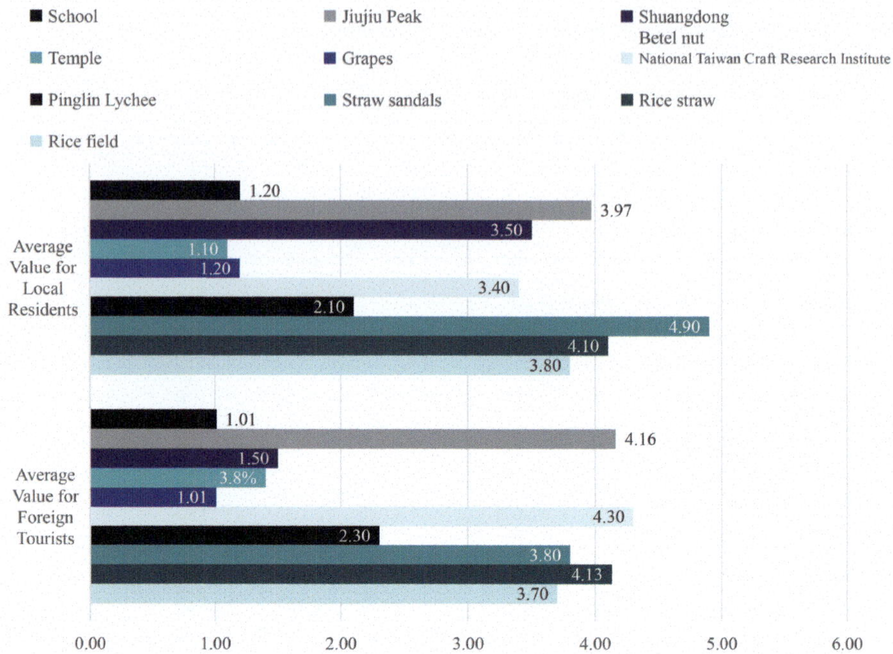

Fig. 5.41 Analysis chart of the opinions of foreign tourists and local residents towards Caotun's spiritual symbols

Table 5.18 Cross tabulation analysis: impression of Caotun's symbols (locals residents vs. foreign tourists)

Option	Identity		
	Average value for foreign tourists	Average value for local residents	Average value of total
Rice field	3.7	3.8	3.75
Rice straw	4.13	4.1	4.12
Straw sandals	3.8	4.9	4.35
Pinglin Lychee	2.3	2.1	2.2
National Taiwan Craft Research Institute	4.3	3.4	3.95
Grapes	1.01	1.2	1.11
Temple	1.4	1.1	1.25
Shuangdong Betel nut	1.5	3.5	2.5
Jiujiu Peak	4.16	3.97	4.07
School	1.01	1.2	1.11

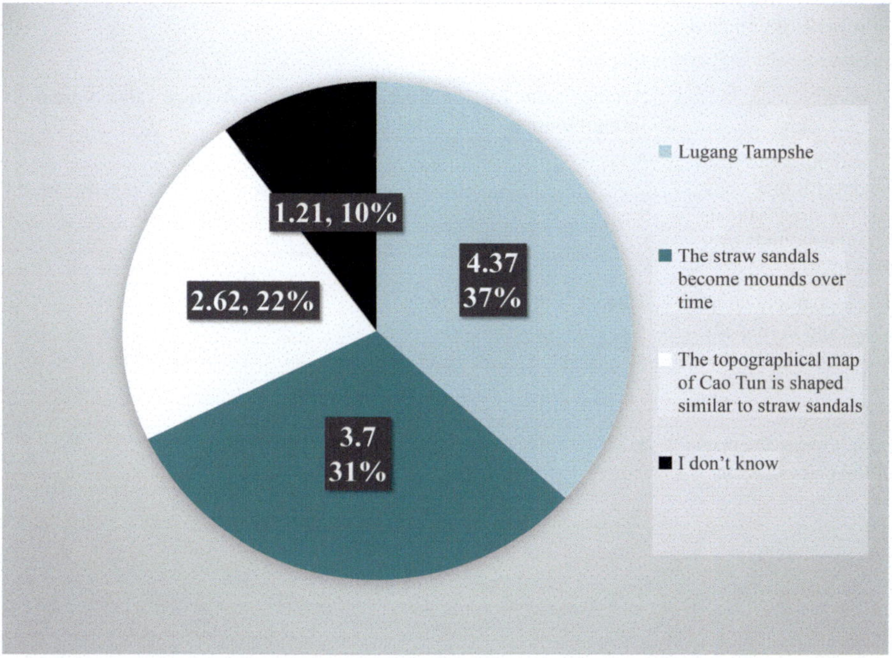

Fig. 5.42 A pie chart analysis of the origin of Caotun's place names as believed foreign tourists and local residents

Table 5.19 Cross tabulation analysis: Caotun's name origin (locals residents vs. foreign tourists)

Option	Identity		
	Average value for foreign tourists	Average value for local residents	Average value of total
Lugang Danpushe, the ancestors went to and from Lugang and Puli, these settlers, porters, and business travelers would arrive in Cao Xie Dun, the land where they would exchange or discard their old straw sandals	3.78	4.96	4.37

(continued)

Table 5.19 (continued)

Option	Identity		
	Average value for foreign tourists	Average value for local residents	Average value of total
Abandoning old straw sandals for new ones, and the straw sandals become mounds over time	3.1	4.3	3.7
The topographical map of Caotun is shaped similar to straw sandals and thus named as straw sandals	1.24	4	2.62
I don't know the origin of the name Caotun	1.22	1.2	1.21

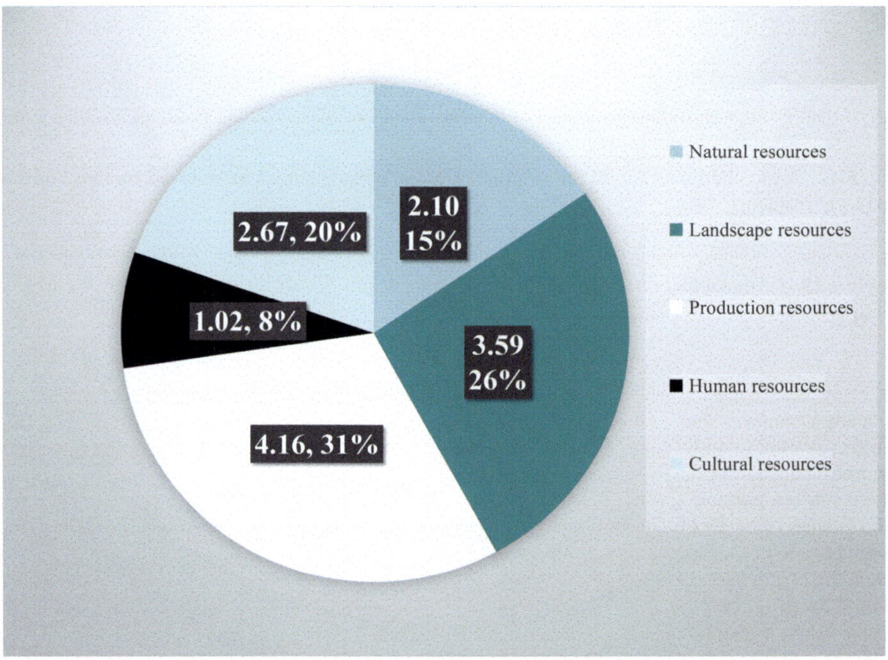

Fig. 5.43 A pie chart analysis of Caotun resources by foreign tourists and local residents

Table 5.20 Cross tabulation analysis: assessment of Caotun's five major resources by locals residents and foreign tourists

Option	Identity		
	Average value for foreign tourists	Average value for local residents	Average value of total
Natural resources	1.56	2.64	2.1
Landscape resources	2.31	4.87	3.59
Production resources	3.92	4.31	4.16
Human resources	1.01	1.03	1.02
Cultural resources	2.23	3.11	2.67

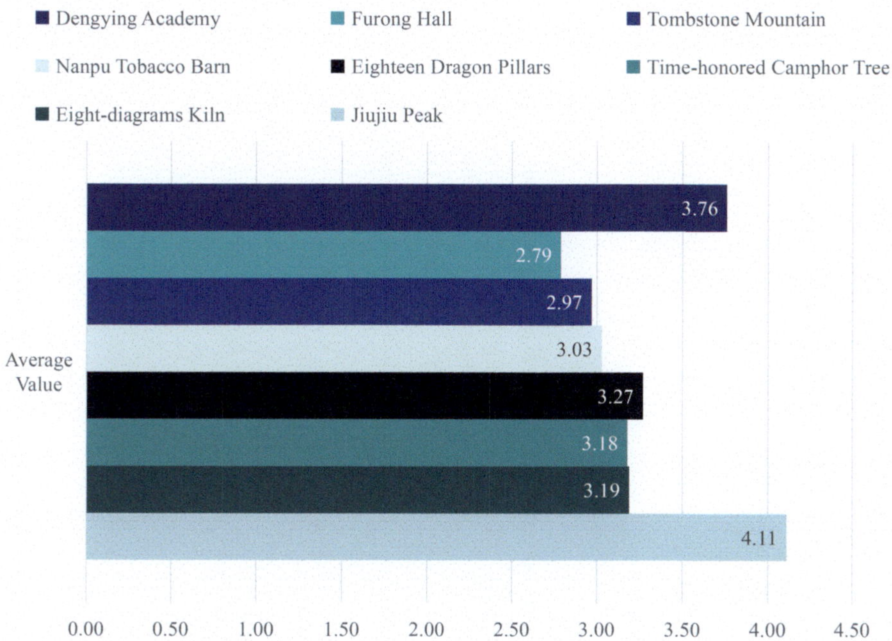

Fig. 5.44 Bar chart analysis of respondents' impression of Caotun's landscape resources

5.5.7 Definition and Status Snalysis of Local Cultural Creative Products

1. What do you think is the definition and status of local cultural creative products among the displayed products?

The main purpose of this question is to understand the respondents' definition of cultural creative products (Table 5.26). As a result, the average value of "products

Table 5.21 Assessment of Caotun's landscape resources by locals residents and foreign tourists

Landscape item	Jiujiu peak	Eight-diagrams Kiln	Time-honored Camphor Tree	Eighteen Dragon Pillars	Nanpu Tobacco Barn	Tombstone Mountain	Furong hall	Dengying academy
Average value	4.11	3.19	3.18	3.27	3.03	2.97	2.79	3.76
Variance	1.01	1.01	1.05	1.28	1.07	1.00	1.07	1.32

Table 5.22 Analysis on purchases of creative merchandise in expos (locals residents vs. foreign tourists)

Question title		Subject		
		Average value for foreign tourists	Average value for local residents	Average value of total
Are you willing to purchase existing local cultural creative products in Caotun Township?	Very willing	1.36	2.78	2.07
	Willing	2.16	1.72	1.94
	Would consider	3.78	2.17	2.98
	Unwilling	2.15	2.57	2.36
	Very unwilling	1.73	3.22	4.95
If the cultural creative products displayed at this exhibition enter the broader market in the future, will you choose to buy them?	Very willing	3.67	4.82	4.25
	Willing	3.14	3.02	3.08
	Would consider	1.34	2.14	1.74
	Unwilling	1.83	1.61	1.72
	Very unwilling	1.01	1.5	1.26

Table 5.23 Feedback on display and planning of creative products from locals residents and foreign tourists

Statistical Methods / Option	Descriptive Analysis		After Factor Analysis		
	Average Evaluation	Variance	Component Matrix After Revolving		
Overall Planning	4.57	.55	.879	.243	-.033
Panel Design	4.46	.61	.820	-.077	.417
Product Design	4.59	.60	.647	.361	.331
Guide Commentary	4.49	.51	.046	.886	.320
Line of Movement Planning	4.51	.51	.300	.780	-.328
Story Application	4.62	.55	.246	.051	.903
Factor Naming			Static Display	Dynamic Commentary	Scenario Guidance
Cumulative Total			33.591	60.107	81.950
Reliability 0.7643					

Table 5.24 Overall evaluation and analysis of creative products expos

Statistical Methods / Option	Descriptive Analysis		After Factor Analysis		
	Average Evaluation	Variance	Component Matrix After Revolving		
Valuable	3.95	0.81	.867	.156	.146
Aesthetic	3.97	0.83	.763	.292	.224
Grace	3.73	0.84	.743	.310	.307
Modernity	3.62	0.85	.717	.297	.224
Connotative	4.00	0.85	.266	.862	.117
Classic	3.97	0.76	.156	.843	-.018
Unique	3.89	0.77	.336	.657	.271
Symbolic	3.89	0.77	.378	.609	.476
Simple	4.00	0.78	.154	.162	.891
Local	4.05	0.74	.259	.029	.747
Practical	3.46	0.87	.407	.551	.576
Factor Naming			Stylistic preference	Cultural understanding	Traditionally imagery
Cumulative Total			28.448	54.682	73.972
Reliability 0.8652					

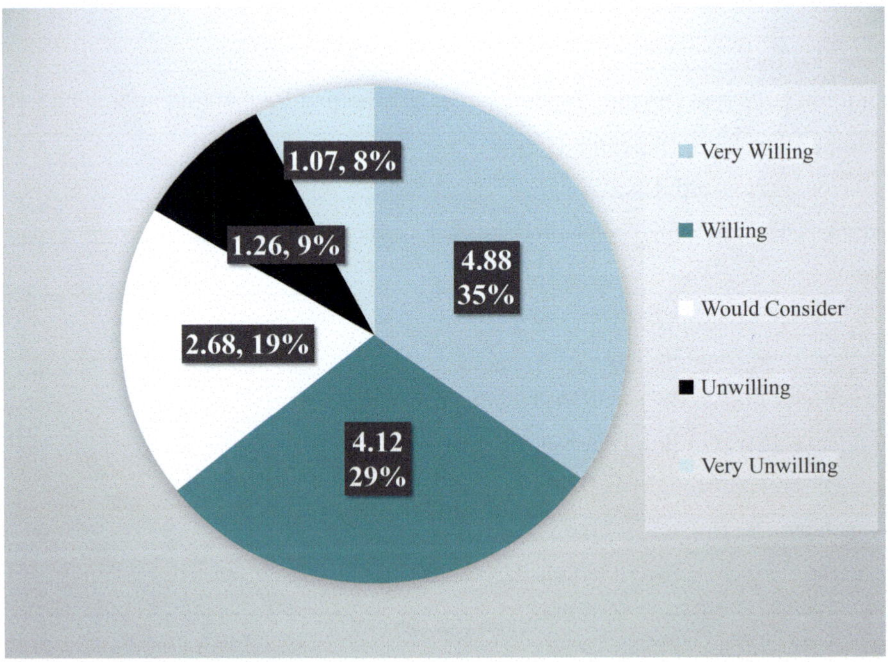

Fig. 5.45 A pie chart analysis of the evaluation of the awareness of Caotun local culture by cultural and creative commodity exhibitions

Table 5.25 Survey on creative products expos' representation of local cultures

Option	Average value
Very willing	4.88
Willing	4.12
Would consider	2.68
Unwilling	1.26
Very unwilling	1.07

Table 5.26 Validity of questionnaires

Questionnaire categories	Identity		Number of Questionnaires	Total
The first category: designers	College department	Cao Xie Dun association	45	415
The second category: regular people	Foreign tourists	Locals residents	370	

with local cultural spirit" is 4.06, with the highest score, The average value of "local, industry, and landscape-related specialty products" is 3.09 and the average value of "products with a rich local story" is 2.27. In addition, this study has a strong understanding of the spiritual factors contained in local cultural creative products, including local products, landscapes and stories, and others which can add value to local cultural creative products. The average value of the option "Commodities between crafts and industrial design" is 1.97, and the average value of the option "Commodities with cultural and commercial value" is 1.39. It shows that local cultural creative products can be produced with less or more, purely handmade, semi-handmade and semi-manufactured, with cultural and spiritual value, decorative functions, stories, history and symbols, and commercial methods such as packaging to better the status of local cultural creative products (Fig. 5.46 and Table 5.27).

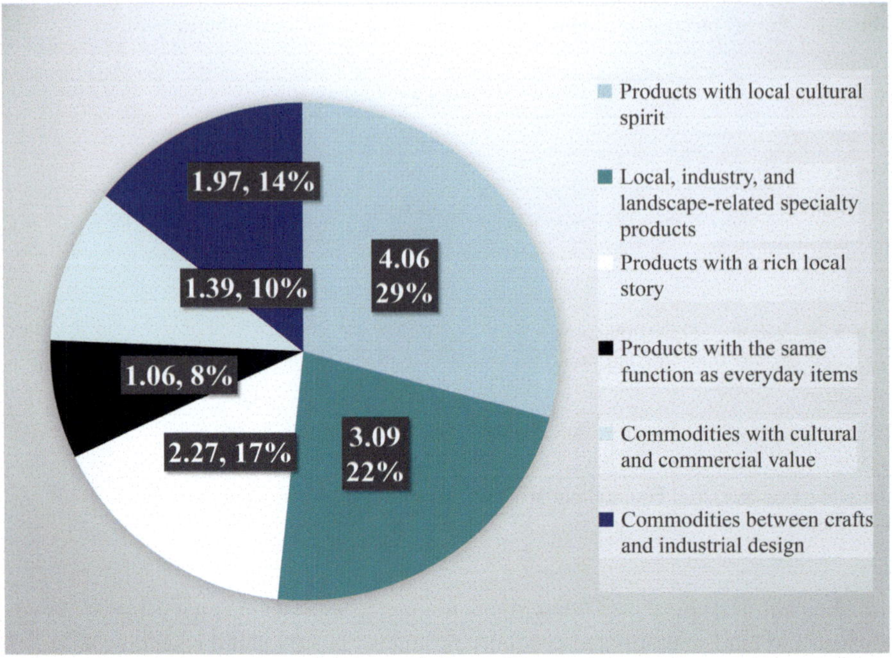

Fig. 5.46 Pie analysis chart of respondents' definition and status of cultural and creative products

Table 5.27 Cross tabulation analysis: positioning of local creative products by designers and consumers

Option	Category				
	Average value for foreign tourists	Average value for local residents	Average Value given by University Department	Average Value given by Cao Xie Dun association	Average value of total
Products with local cultural spirit	3.85	3.91	4.82	3.67	4.06
Local, industry, and landscape-related specialty products	1.86	3.96	4.94	1.6	3.09
Products with a rich local story	1.01	2.35	3.11	2.59	2.27
Products with the same function as everyday items	1.02	1.05	1.08	1.07	1.06

(continued)

Table 5.27 (continued)

Option	Category				
	Average value for foreign tourists	Average value for local residents	Average Value given by University Department	Average Value given by Cao Xie Dun association	Average value of total
Commodities with cultural and commercial value	1.04	2.35	1.07	1.11	1.39
Commodities between crafts and industrial design	2.57	3.21	1.06	1.03	1.97

References

Compiled by Caotun Town Office. (1986). *Caotun township chronicle*. The Caotun Township Office, Nantou County.

Compiled by Caotun Town Office. (2005). *Continued edit of the caotun town chronicle*. The Caotun Township Office, Nantou County.

Gabriel, Y. (1998). The use of stories. In G. Symon & C. Cassell (Eds.), *Qualitative Methods and Analysis in Organizational Research*. Sage Publications.

Green, W. S., & Jordan, P. W. (2002). *Pleasure with Products: Beyond Usability*. Taylor & Francis.

Huang, F. X. (2001). *Discussion on the spontaneity of local residents and the revitalization of local traditional crafts—using yingge ceramics as an example*, Master's thesis. Department of Industrial Design, Tunghai University, Taichung.

Jarke, M., Tung, B. X., & Carroll, J. M. (1998). Scenario management: An interdisciplinary approach. *Requirements Engineering, 3*(3–4), 155–173. https://doi.org/10.1007/s007660050002

Kuo, P. H. (2016). Effects of synchronous web-based instruction on students' thinking styles and creativity. *Eurasia Journal of Mathematics, Science & Technology Education, 12*(3), 609–619.

Makela, A., & Mattelmaki, T. (2002). Collecting Stories on user experiences to inspire design—A Pilot. In W. S. Green & P. W. Jordan (Eds.), *Pleasure with Products: Beyond Usability* (pp. 333–344). Taylor & Francis.

Moggridge, B. (1993). Design by story-telling. *Applied Ergonomics, 24*(1), 15–18.

Pine II, B., & Gilmore, J. (2013). *Chapter 2: The experience economy: past, present and future* (pp. 21–44). UK: Edward Elgar. https://doi.org/10.4337/9781781004227.00007

The Beautiful Home Editorial Department. (2019). *The Display Design Bible*. My House Publishing Co., Ltd.

Tseng, Y. L. (2002). *The application of protection design to the design of computer bag: By using scenario design and storytelling approach*. Master's thesis. Department of Industrial Design, Chang Gung University, Taipei.

Yu, D. Z., Lin, W. Q., & Wang, J. Q. (2000). *Scenario-Oriented Design: New Law of Product and Service Design in the Information Age*. Garden City Publishing House.

Chapter 6
Comprehensive Discussion

Based on the application of the knowledge base and the statistical basis of the display effect of cultural creative products, the following section will discuss the construction process of the conceptual form of the knowledge base and the comprehensive analysis results of the scenario-based cultural creative product exhibition.

6.1 Analysis Results and Discussion of Knowledge Base Conceptual Form Establishment and Process

1. Knowledge base effectiveness
(1) The research in this book proves that the knowledge base is the best bridge between "foreign designers" and "local designers"; because it enables "foreign designers" to understand the local historical background and cultural status, while it is helpful for "local designers" as it assists them in their efforts to understand the product development process and knowledge base shape, color, texture, material and cultural connotation, and then convert it into design elements (Fig. 6.1).
(2) The conceptual model of the knowledge base in this study, after verification by designers in other places and localities, has been statistically proven to be applicable to the cases of cultural creative products in other local industries.
2. Analysis of the Advantages and Disadvantages of Using the Knowledge base for Local and Foreign Designers
(1) The advantages of the Cao Xie Dun Association: members of the Cao Xie Dun Association have a higher understanding of local history and stories; compared with the foreign designers of the university department, they have the advantage. The Cao Xie Dun Association believes that "final cultural creative product model making" is easiest, because the local residents have been deeply involved in the technical learning of handicrafts over the years.

M.-F. Wang, *Creative Product Design With Cultural Codes*,
https://doi.org/10.1007/978-981-19-6767-2_6

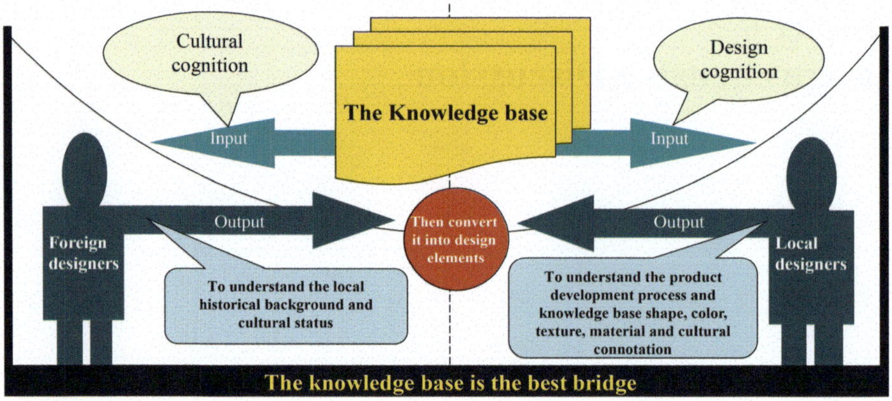

Fig. 6.1 Knowledge base is the bridge between foreign and local designers

(2) Disadvantages of the Cao Xie Dun Association: "Brainstorm/Idea Sketch" is the most difficult for members of the Cao Xie Dun Association. Because local residents have no professional training in design, oftentimes ideas cannot be presented during the development of concept sketches, and concepts cannot be implemented into physical products.
(3) Advantages of the university department: In terms of "Brainstorm/Idea Sketch," members of the University Department are far more apt when compared to members of the Cao Xie Dun Association, this is because all members have been trained professionally during their time in university.
(4) Disadvantages of the university department: The awareness of cultural background is not enough, but it can narrow the difference in the awareness of local industries through the sorting of local cultural data in the knowledge base. In addition, due to the low level of product refinement in the university department, the finished product cannot truly convey the local cultural image and code.

After the above statistical analysis of the feasibility of using the knowledge base, it is verified that the knowledge base, when used in the development mode of cultural creative products, is helpful for product designers. However, finished products must still be highly evaluated by foreign tourists and local residents. In the future, the cultural creative product exhibition will be planned by the scenario-based design method to further understand the perception and evaluation of cultural creative products by foreign tourists and local residents.

6.2 Analysis Results and Discussion of the Scenario-Based Cultural Creative Product Exhibition

1. Subjects' goals for the commercialization of culture creative industries

The data shows the public's demand for the consumption of local cultural creative products, and the local products show a unique spirit, which improves the testees' goal of commercializing local cultural creative products. However, the lack of sophistication of cultural creative products reduces the possibility of commercialization.

2. Subjects' perception of culture creative commodity exhibitions

After seeing the exhibition, the subjects gave high evaluations of Caotun culture. The data shows that watching the exhibition, walking the map, telling stories, and understanding the culture show and performance methods can improve the awareness of Caotun culture.

6. The effect of introducing the scenario-based method into the exhibition

The data shows that the overall planning of the exhibition site is highly rated. This displays the effectiveness of the scenario-based exhibition in strengthening the story. If the style of the exhibition products can be unified or consistent, the exhibition will be more complete.

7. Cultural impression and understanding of Caotun

Regarding the foreign tourists and local residents, with the background of "Lugang Danpushe," the scenario-based design method was written to increase the awareness of the exhibition. It will also affect the culture and impression of Caotun because of the choice of local industrial, natural, human, cultural, and scenic resources.

8. Definition and status of local cultural creative products

1. The respondents' definition of local cultural creative products

The respondents included tourists from other places, local residents, university departments and the Cao Xie Dun association, among others. For the options "products with local cultural spirit" ranked first, "products with strong local stories" ranked second, and "products related to places, products, and landscapes" ranked third. This data shows that the basic conditions for the formation of local cultural creative products are the local cultural spirit, stories, products and landscape elements. This data is the preferred choice of the respondents in this study, and it also proves that the spiritual connotation of local cultural creative products is the most important (Fig. 6.2).

2. Subjects' status of local cultural creative products

The average value of selecting "products between craftsmanship and product design" is the highest, which further indicates that the respondents have different orientations towards the three types of markets, including local cultural creative products, craftsmanship and product design, for example: "craftsmanship" attaches importance to small-scale production, purely hand-made, products which feel good in the hand, and the function tends to be purely decorative. "Product Design" attaches great importance to mass production, human factors considerations, market-oriented consumer

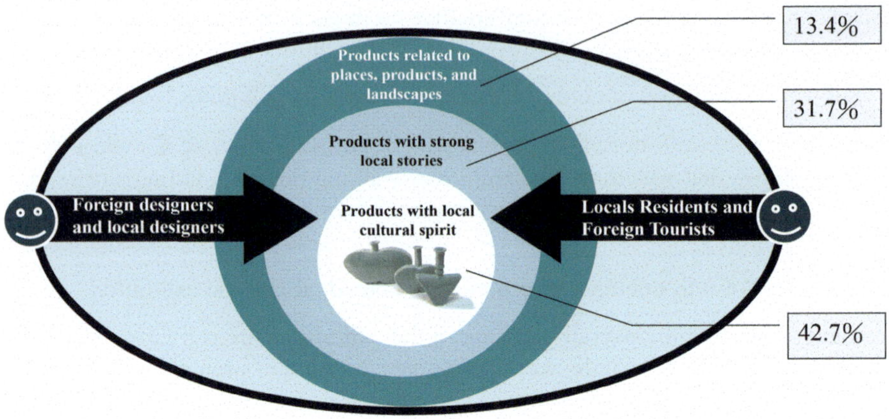

Fig. 6.2 The respondents' definition of local cultural creative products

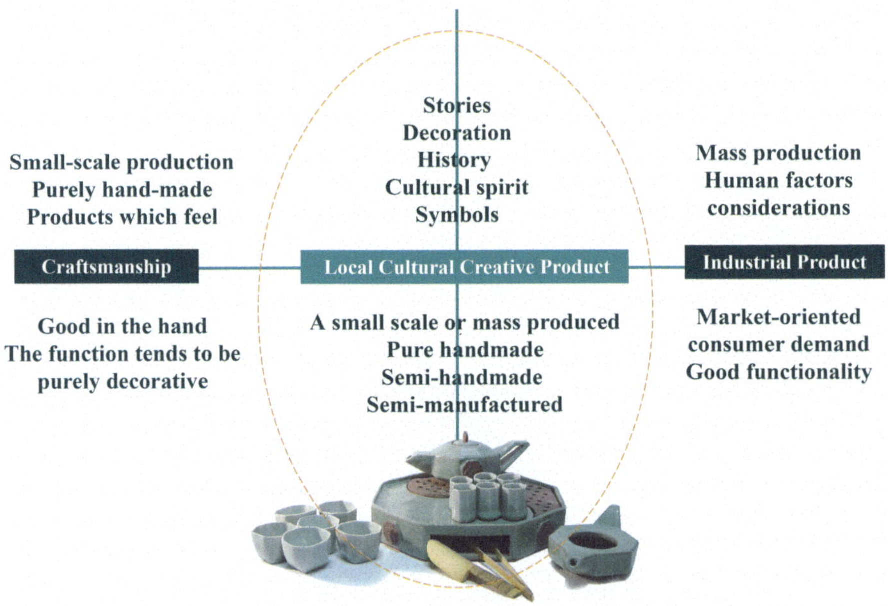

Fig. 6.3 The respondents' determination of the status of local cultural creative products

demand and good functionality. Its status as a local cultural creative product, the results of this research are biased towards products which could be both produced on a small scale and mass produced, pure handmade, semi-handmade semi-manufactured, cultural spirit, decoration, stories, history and symbols, etc. (Fig. 6.3).

Chapter 7
Conclusions and Recommendations

This book mainly explored the creative product design of the cultural codes, in the conclusion and suggestions section, we will explore users of the design model and the consumers of the cultural creative products exhibition, expounding comprehensively the verification results. To propose possible directions of continued research in the future.

7.1 Summary of the Book

1. Verification results and questionnaire regarding the feasibility of implementing the knowledge base design model

(1) To reduce designers' cognition differences of cultural and creative products

It is confirmed that designers' cognition of local culture can be raised gradually so that the time for acquiring local information can be shortened; the local designers have their own standards on the cognition of local culture, but they lack the designing professional knowledge. After the experiment of creating cultural and creative products with the knowledge base, it is believed that extracting such important elements for designing as shape, characters, symbols can help designers gradually get into the state and immerse themselves in the designing in the future.

(2) The simple and convenient designing model enables the designers to develop cultural and creative products independently

From the miniature framework of the knowledge base to the use of charts and steps in its content and form, it can be seen from the theoretical examples of semiotician that it is not only easy to understand and distinct to read, but also convenient for the designers to manipulate.

M.-F. Wang, *Creative Product Design With Cultural Codes*,
https://doi.org/10.1007/978-981-19-6767-2_7

(3) To reduce designers' cognition differences of cultural and creative products

It is confirmed that designers' cognition of local culture can be raised gradually so that the time for acquiring local information can be shortened; the local designers have their own standards on the cognition of local culture, but they lack the designing professional knowledge. After the experiment of creating cultural and creative products with the knowledge base, it is believed that extracting such important elements for designing as shape, characters, symbols can help designers gradually get into the state and immerse themselves in the designing in the future.

(4) Design models can assist non-professional designers in developing products

Most of the local cultural creative product designers do not have professional design experience. After being trained in the study design model used by this study, and the completion of the experiment, it was noticed that local cultural creative product designers have a marked difference in their knowledge of designing, including the concept of design, the shape composition, and thinking three-dimensionally. Thus, proving that the cultural creative product design model used in this study, can assist non-professional designers in developing products.

(5) The results of the designer's learning of the cultural creative product development model with different cultural understanding

After training on the development model of cultural creative products used in this study, it is found that local and non-local designers have different perceptions of design products and culture before designing. However, the new products developed by designers after training on using the knowledge base to design cultural creative products were far different from those that they had developed previously. For example, after deepening the impression of the local culture, they can extract local symbolic symbols through the design profession, which can be effectively used in cultural creative products and present special products with more local connotations.

2. Evaluations of the understanding/cognition levels regarding the scenario-based cultural creative product exhibit

(1) To narrow the differences in the subjects' understanding of local cultural creative products

The research and study found that planning the cultural creative product exhibition with the scenario-oriented design method enables local residents to have a preliminary understanding of local culture. After their participation in the cultural creative product exhibition, For the part of foreign tourists, they deepened their feelings for the product exhibitions and stories extended by the local industry, shrinking the understanding gap. Even though in the cultural creative product exhibition, the function of guided tours is added, even if foreign tourists and local residents choose to not participate in guided tours, their evaluation of their understanding remains high.

(2) The consumption behavior and motivations of subjects towards local cultural creative products

In the questionnaire survey results of the three cultural creative product exhibitions, both foreign tourists and local residents have improved purchasing motives for cultural creative products. The respondents also asked many times at the exhibition site until the completion of the questionnaire whether the exhibition goods are available for purchase. Thus showing that both mediums, from the questionnaire results to the physical products at the exhibition site, can give consumers the motivation and desire to purchase.

(3) The substantial results of the scenario-based design method

The scenario-based design method guides the cultural creative product exhibition, and it can be found that the static display is integrated into the explanation of the tour guide, which can improve the subject's understanding of local culture; without the explanation of the tour guide, the subject's local cultural understanding remains relatively high. The reason for the improvement is that the exhibition introduces the scenario-based design method, and integrates the local stories, situations and culture of Caotun into the exhibition, so that the subjects can experience the same cultural background in the same space environment, and achieve a static exhibition with the ability to achieve the overall effect of "see the exhibition, walk the map, tell stories and understand the culture."

3. The combination effect of building a knowledge base with a theoretical foundation based in semiotics and the scenario-based cultural creative product exhibition

After verifying the feasibility of implementing the knowledge base design model through the first test, the products developed through the knowledge base are used to guide the cultural creative product exhibition. A second test was carried out at the same time to confirm that consumers will purchase the products. This two-stage procedure, method, and results achieve the overall structure of the local cultural creative product design model in this book, the development model not only verifies the development and application results of designers, but also extends to consumers' preference for the knowledge base development model and the created cultural creative products.

4. Results of integrating the industrial design and development process into the design process of cultural creative products

The industrial design and development process has the advantage of being rigorous and reviewing the details and quantification of each stage of development. This process would aid in promoting the commercialization of cultural creative products in local industries. In this study, it is proved that the products developed by designers can indeed make most of cultural creative products achieve the goal of "commercialization of cultural creative products; culturalization of commercial products."

7.2 Recommendations of the Book

1. The reference value of the design model used in this book in terms of its applications to the development of cultural creative products in other industries in other localities

The local cultural creative product design mode is a shortcut to simplify huge and complex local resources, integrate the conceptual form of the local cultural and historical design symbol knowledge base, and assist foreign designers to understand local culture. Allowing designers with a lower skill level to have easy access to design elements to incorporate into merchandise. The effectiveness of the design model constructed in this book is helpful for local industries to be used as a reference, or to continue research and development.

2. The effectiveness of the scenario-based cultural creative product exhibition

As a result of this study, the exhibition uses the scenario-based method to introduce the background story of the Lugang Danpushe, so that foreign tourists and local residents can increase their awareness of Caotun culture during the exhibition, and at the same time give public a deeper understanding of the cultural creative product exhibition. This method can be used in future research to explore cultural centers, art galleries and museums with local characteristics, so as to prove the influence of the scenario-based method on curating exhibitions and planning.

3. Using the book to organize local resources

This book classifies and studies the industrial resources of Natural, Production, Human, Cultural and Landscape, and effectively achieves the basic background information required for the design of cultural creative products. This publication uses the resources invested in research to activate new resources for local industries.

4. Sustainable innovation

Although this study proposes the concept of a local cultural and historical design symbol knowledge base to solve the problems faced by the design and promotion of cultural creative products, local cultural creative product designers must continue to innovate, and can continue to review and improve cultural creative products. On the one hand, it can maintain the appeal of new products towards tourists and consumers, and on the other hand, it can differentiate products from similar industries, and sustain the economic market of local industries.

5. Further the development of a digital knowledge base

In the era of big data, the knowledge base will naturally become more digitized along with the integration of technology into local culture industries. Allowing for the knowledge base to be used and searched from more easily, making the input and replacement of local resources clearer and more convenient.

6. Follow-up discussion on the widespread use of the knowledge base of cultural and historical design symbols

In the future, the promotion of local cultural industries in Taiwan will face huge industrial resources. We will use the knowledge base to integrate and use the aspects of use in local construction, overall community development, cultural tours, local public art, cultural festivals, local creation, software and hardware planning and cultural commodity design, among others, to promote the revitalization of local industries. Although this book uses the knowledge base in the local cultural creative product design program, when conducting research, the use of the knowledge base can be extended to related industries in various localities, providing an additional reference for Taiwan's culture creative industries.

Index

B

Bangka, 112, 148
Beitou, 7, 62, 68, 109, 110, 112, 146, 148, 154, 157

C

Caotun Township, 3, 4, 7, 8, 27, 29, 46, 61, 62, 64, 68, 69, 72, 74, 90, 92, 125, 148, 167, 177, 197
Cao Xie Dun, 7–9, 16, 17, 19, 21, 27–29, 61–64, 67–70, 92, 110, 111, 121, 125–127, 131, 134, 135, 146, 148, 154, 174, 177, 182, 184, 193, 199–201, 203–205
Cao Xie Dun Local Cultural and Educational Association (Cao Xie Dun Association), The, 17, 126
Changhua, 7, 62, 109, 110, 146, 148, 154, 155
Community Empowerment/Development, 1, 3, 5, 31–33, 35, 38, 52, 85, 211
Cultural and creative products, 4, 6, 8–10, 13, 16–18, 20, 21, 25, 28–31, 38–41, 43–46, 51, 55, 57, 70, 74, 76–79, 84, 85, 89, 99, 100, 103, 116–118, 125, 153, 200, 207, 208
Cultural and creative products design, 8
Cultural and creative products design and development model, 6, 89, 99
Cultural and creative products exhibition, 145
Cultural and historical design symbols knowledge base, 4, 6, 98, 114–118
Cultural concreteness, 4, 6, 104, 105, 114, 118, 121, 136, 164, 168, 184, 185
Cultural connotation, 2, 4, 6, 29, 36, 76, 84, 103–105, 114, 118, 121, 137, 161, 168, 171, 174, 184, 190, 203
Culture, 1–5, 8, 9, 16–18, 20, 25–41, 52, 55, 57, 61, 72, 74, 77, 78, 81, 85, 89, 93, 95, 100, 103–105, 108, 120, 121, 125, 126, 141–143, 145, 146, 148, 149, 152, 156, 166, 169, 170, 185–189, 191, 198, 199, 205, 207–210
Culture creative industries, 4, 5, 204, 211
Culture industries, 2–6, 27, 32, 34, 43, 210
Culture's intermediate, The, 104, 105
Culture's metaphysical, The, 104
Culture's physical, The, 104

D

Dengying Academy, 27, 28, 95, 107, 111, 113, 154, 155, 157, 165, 168, 178, 180, 196
Department of Industrial Design, Tunghai University (The University Department), 9, 21, 125, 126

E

Eight-diagrams Kiln, 16, 27, 28, 107, 112, 113, 154, 155, 159, 175, 177, 178, 180

F

Field survey, 15, 25, 27, 28, 106, 113
Foreign designers, 4, 8, 10, 17, 21, 125, 126, 131, 203, 210

M.-F. Wang, *Creative Product Design With Cultural Codes*,
https://doi.org/10.1007/978-981-19-6767-2

Foreign tourists, 8, 9, 12, 20, 76, 85, 141, 165, 166, 169, 172, 174, 176–178, 186, 192–197, 199–201, 204, 205, 208–210
Form, Color, Material, Texture, Function and Spirit, 6, 47, 51, 82, 84, 100–104, 114, 116–118, 120, 134–137, 203
Furong Hall, 16, 27, 28, 107, 109, 111, 113, 154, 155, 158, 169, 171, 180

I
Indoor healthy straw slippers, 72, 97, 174
Item Presentation (IP), 147, 152

J
Jiujiu Peak, 16, 27, 28, 100, 106–108, 113–118, 120, 121, 155, 161, 179, 180, 183–185, 192, 196

L
Local cultural and creative products, 3, 4, 6, 9, 10, 17, 20, 25, 29, 39, 45, 46, 69, 72, 74, 76, 85, 89, 99, 102, 105
Local cultural industry, 3, 25, 31, 43, 74, 143
Local designers, 8, 9, 13, 17, 19–21, 84, 85, 116, 121, 125, 126, 131, 141, 203, 204, 207, 208
Locals residents, 192, 193, 195–197, 199
Lugang, 7, 61, 62, 69, 107, 145, 146, 148, 154, 155, 157, 159, 161–163, 177, 193
Lugang Danpushe, 4, 7, 27, 28, 90–92, 107, 110, 113, 142–144, 146, 148, 149, 153, 155, 161–164, 177, 178, 190, 193, 205, 210

N
Nanpu Tobacco Barn, 154, 155, 160, 178, 180, 196
National Palace Museum, 43, 44, 46–51, 53, 74
National Taiwan Craft Research Institute, 112, 192
Natural, Production, Human, Cultural and Landscape, 8, 25–28, 100, 107, 108, 114, 117, 148, 149, 178, 210

P
Paradigmatic, 91–93, 102, 116
Point of Sales Presentation (PP), 147, 149
Pragmatic, 93, 102, 103
Puli, 7, 61–64, 69, 70, 107, 110, 111, 146, 148, 154, 155, 158, 162, 177, 188, 190, 193

Q
Qualitative research, 4, 5, 13–16, 22
Quantitative analysis, 4, 5, 19, 165, 178

R
Regional revitalization, 211

S
Scenario-oriented design, 4, 10, 13, 15, 18–20, 85, 141, 165, 178, 208
Semantic, 83, 92, 102, 103
Semiology, 18, 103
Signified, 18, 90–92, 94, 102, 119
Signifier, 18, 90–92, 94, 102, 119
Statistics Package for the Social Sciences (SPSS), 20
Straw Sandals, 7, 16, 27, 28, 36, 61–63, 69, 70, 72, 73, 90–92, 97, 98, 107, 110, 111, 113, 144, 148, 154, 155, 158, 159, 164, 173, 174, 176–178, 190, 192–194
Syntagmatic, 91–93, 102, 116
Syntax, 92, 93

T
Tainan, 2, 7, 62, 148
Time-honored Camphor Tree, 106, 109, 113, 178, 196
Tombstone Mountain, 27, 28, 106–108, 113, 154, 155, 161, 162, 186, 188, 196

V
Visual Presentation (VP), 147, 148

W
Wanhua, 7, 62, 148

The manufacturer's authorised representative in the EU is Springer Nature Customer Service Centre GmbH, Europaplatz 3, 69115 Heidelberg, Germany. If you have any concerns regarding our products, please contact ProductSafety@springernature.com

Printed and bound by CPI Group (UK) Ltd, Croydon, CR0 4YY
30/09/2026
02225317-0006